DFT Based Studies on Bioactive Molecules

Authored by

Ambrish Kumar Srivastava
Department of Physics
Deen Dayal Upadhyaya Gorakhpur University
Gorakhpur
Uttar Pradesh
India

&

Neeraj Misra
Department of Physics
University of Lucknow
Lucknow Uttar
Pradesh
India

DFT Based Studies on Bioactive Molecules

Authors: Ambrish Kumar Srivastava and Neeraj Misra

ISBN (Online): 978-981-4998-36-9

ISBN (Print): 978-981-4998-37-6

ISBN (Paperback): 978-981-4998-38-3

©2021, Bentham Books imprint.

Published by Bentham Science Publishers Pte. Ltd. Singapore. All Rights Reserved.

need for a court order if at any point you breach any terms of this License Agreement. In no event will any delay or failure by Bentham Science Publishers in enforcing your compliance with this License Agreement constitute a waiver of any of its rights.

3. You acknowledge that you have read this License Agreement, and agree to be bound by its terms and conditions. To the extent that any other terms and conditions presented on any website of Bentham Science Publishers conflict with, or are inconsistent with, the terms and conditions set out in this License Agreement, you acknowledge that the terms and conditions set out in this License Agreement shall prevail.

Bentham Science Publishers Pte. Ltd.
80 Robinson Road #02-00
Singapore 068898
Singapore
Email: subscriptions@benthamscience.net

BENTHAM SCIENCE

CONTENTS

FOREWORD

Density functional theory (DFT) based methods use the description of the electronic density of an atom or molecule to calculate a host of important properties, many of which are not easily obtained via experimental methods. Some attributes that may be accessed in this manner include molecular geometry, vibrational frequencies, dipole moments and higher-order moments, thermochemical properties, and so forth.

It gives me great pleasure in writing the foreword of this book. It is an outcome of a rigorous amount of effort, which has been devoted to conceptualizing, planning, and finally writing the book. The book contains all the ingredients required to understand, practice, and perform the DFT based studies. The first chapter of the book introduces the concept of DFT and the second chapter deals with its application to explore molecular systems using the popular *Gaussian* program. The subsequent chapters of the book discuss the results obtained by DFT calculations of various biologically important molecules. The last chapter exclusively focuses on the quantum theory of atoms in molecules, used for the study of various inter- and intra-molecular interactions. The book is also complemented with a sample output of the *Gaussian* as an appendix, which can be used to extract and interpret the results of DFT based calculations.

Despite the availability of high-performance computing and the emergence of new theoretical approaches, understanding of structure↔function correlation in molecular and macromolecular systems remains an elusive goal. I am confident that this book shall be of immense value for students, young researchers, scientists, teachers, and all those interested in exploiting DFT methods for molecular systems, particularly biologically active compounds. This book will help to learn and master the technique of applying the DFT based methods and the *Gaussian* program for analyzing various properties of biologically active molecules.

With best wishes,

Sugriva Nath Tiwari
Dean, Faculty of Science, Professor and Former Head, Department of Physics
Deen Dayal Upadhyaya Gorakhpur University, Gorakhpur, India
& Former President, Indian Science Congress Association (Physical Sciences Section)

PREFACE

The very idea of writing a book on density functional theory (DFT) based studies on molecular systems arose from the volume of work carried out by us over a while. We have always felt the need for a concise literature on the theory and practice of DFT followed by a proper compilation of the research work using the well-known suite of programs, such as, the *Gaussian*. The sole perspective of initiating this project was to make available a good pool of literature, which can presumably be of immense help to the young researchers and experimentalists among others, who are planning to work or have been already working in this rapidly growing and exciting field of research.

The book has been organized into seven chapters and written from the beginners' perspective in such a way that anyone interested to work on molecular systems using the DFT based methods and the *Gaussian* program, can get an exhaustive and a very apropos idea of "how to employ the DFT on molecules" to explore the various properties of the systems under study. The chapters of the book have been methodically presented so that before starting to work on any molecular system, it is assumed that the reader gets well acquainted with the basics of DFT. After becoming friendly with the fundamentals of DFT, the reader is exposed to the applications of DFT on molecular systems with the focus on the *Gaussian* and its usage in a much applied way. Thereafter, many interesting themes have been covered in the form of the subsequent chapters of the book, namely, DFT studies on synthetic compounds, unusual amino acids, and natural products followed by a chapter on a comprehensive account on the way theory is used to complement the experiment. Considering the role of interactions in biologically active molecules, an exclusive chapter on the quantum theory of atoms in molecule (QTAIM) has been included. To supplement the second chapter and make the content more digestive, an appendix has also been added.

All in all, we tried every effort to present a concise and at the same time, complete picture of DFT and its role, action, and applications on some biologically active molecules. We believe that this book will serve its purpose and all the readers, irrespective of their field and level of experience would benefit in some way or the other.

We wish you a happy DFT.

Ambrish Kumar Srivastava
Department of Physics Deen Dayal Upadhyaya Gorakhpur University
Gorakhpur, Uttar Pradesh
India

&

Neeraj Misra
Department of Physics University of Lucknow
Lucknow, Uttar Pradesh
India

CHAPTER 1

The Essence of Density Functional Theory

Abstract: This chapter outlines the basic principles of the density functional theory (DFT). The introduction of electron density to develop the Kohn-Sham approach has been systematically presented. The various approximations such as LDA, GGA, and hybrid functional for the exchange-correlation energy have been discussed. A separate discussion on the basis sets has also been included. The advantages and shortcomings of DFT based techniques are also revealed. The formulation of time-dependent DFT has been presented in a concise manner. This chapter is intended to provide an overview of the theoretical background of the methods adopted in the succeeding chapters.

Keywords: Basis sets, DFT, Electron density, Exchange-correlation energy, Gaussian, Generalized-gradient approximation, Gradient-corrected functional, Hohenberg-Kohn theorem, Hybrid functional, Kohn-Sham approach, Local density approximation, TDDFT.

INTRODUCTION

The central idea behind the density functional theory (DFT) is a different variant of quantum mechanics, and like the wavefunction-based methods, some DFT methods do not use any empirical parameters and are derived from the first principles. In contrast to wavefunction-based methods, however, instead of using approximate molecular orbital wavefunctions, DFT uses the knowledge of the overall electron density to solve for the desired properties. Methods based on DFT have gained in popularity due to recent theoretical advancements that often allow it to achieve greater accuracy, at a lower or similar cost in computation time, than commonly used wavefunction-based methods such as the Hartree-Fock (HF) theory.

Since the book is intended for the application of density functional theory (DFT) and time-dependent DFT methods, it is very relevant to describe the formulation of the theory.

The Schrödinger Equation

DFT attempt to solve the non-relativistic Schrödinger wave equation:

Ambrish Kumar Srivastava and Neeraj Misra

$$\hat{H}\Psi = E\Psi \tag{1}$$

Here Ψ is the wavefunction, $\hat{H}$ is time-independent non-relativistic Hamiltonian, and E is the energy of the system.

$$\hat{H} = \hat{T} + \hat{V} \tag{2}$$

The kinetic energy operator $\hat{T}$ can be expanded into the following components:

$$\hat{T} = -\sum_{i=1}^{N} \frac{1}{2}\nabla_i^2 - \sum_{A=1}^{M} \frac{1}{2M_A}\nabla_A^2 \tag{3}$$

where the first term is the kinetic energy for the electrons and the second is that for the nuclei. Similarly, the potential energy operator ($\hat{V}$) is given by.

$$\hat{V} = -\sum_{A=1}^{M}\sum_{i=1}^{N}\frac{Z_A}{r_{iA}} + \sum_{i=1}^{N}\sum_{j>i}^{N}\frac{1}{r_{ij}} + \sum_{A=1}^{M}\sum_{B>A}^{M}\frac{Z_A Z_B}{R_{AB}} \tag{4}$$

Here the first, second, and third terms represent the electron-nucleus attraction, the electron-electron repulsion, and the nucleus-nucleus repulsion, respectively.

Needless to mention that the Schrödinger equation can't be solved "exactly" for any system other than the simplest (single-electron) atomic system [1]. To solve this, therefore, we require certain approximations as discussed below.

Born-Oppenheimer Approximation

The complete non-relativistic Hamiltonian using eq. (2), (3) and (4) is given below,

$$\hat{H} = -\sum_{i=1}^{N}\frac{1}{2}\nabla_i^2 - \sum_{A=1}^{N}\frac{1}{2M_A}\nabla_A^2 - \sum_{A=1}^{M}\sum_{i=1}^{N}\frac{Z_A}{r_{iA}} + \sum_{i=1}^{N}\sum_{j>1}^{N}\frac{1}{r_{ij}} + \sum_{A=1}^{M}\sum_{B>A}^{M}\frac{Z_A Z_B}{R_{AB}} \tag{5}$$

One can write $\hat{H}$ into two parts considering the nuclear and electronic motions separately,

$$\hat{H}_{elec} = -\sum_{i=1}^{N} \frac{1}{2}\nabla_i^2 - \sum_{A=1}^{M}\sum_{i=1}^{N}\frac{Z_A}{r_{iA}} + \sum_{i=1}^{N}\sum_{j>i}^{N}\frac{1}{r_{ij}} \qquad (6)$$

$$\hat{H}_{nucl} = -\sum_{A=1}^{M}\frac{1}{2M}\nabla_A^2 + \sum_{A=1}^{M}\sum_{B>A}^{M}\frac{Z_A Z_B}{R_{AB}} \qquad (7)$$

According to the Born-Oppenheimer approximation, the motion of the electrons in a molecule can be considered in a field of fixed nuclei. This is based on the fact the nuclei are much heavier than the electrons. This implies that the kinetic energy of the nuclei, the first term in eq. (7) can be neglected and the nuclear repulsion energy, the second term in eq. (7), becomes constant for a specific molecular geometry [2]. Therefore, one has to deal with the electronic Hamiltonian, eq. (6). The eq. (6) can be solved for the electronic energy ($E_{elec.}$) considering a fixed set of nuclear coordinates. The total energy is then simply a sum of E_{elec} and the constant nuclear repulsion energy.

Electron Density and Wavefunction

Note that the electronic wavefunction (ψ) obtained by solving eq. (6) is not measurable or observable. The experiments can measure several parameters of molecular systems, including electron density (ρ), which is measurable by X-ray diffraction or electron diffraction. It might be a great idea to use one-electron density instead of many-electron wavefunctions for calculating the molecular geometries, energies, *etc*. In Table **1**, we compare the properties of wavefunction and electron density.

Table 1. Comparison of electron density and wavefunction.

Electron density	Wavefunction
Observable	Not observable
Real	Complex
One electron	Many electrons
3 coordinates for N-electron systems	$3N$ coordinates for N-electron systems

In the Born interpretation, the probability density at any point is nothing but the one-electron wavefunction (ψ) squared (having the same unit as that of the wavefunction at that point). For multi-electron wavefunction, the relation between ρ and ψ is more complicated. Nevertheless, the relation between ρ and ψ reads,

$$\rho = \sum_{i=1}^{N} n_i |\psi_i|^2 \tag{8}$$

where ψ_i is the one-electron spatial wavefunctions.

THE KOHN-SHAM APPROACH

The Kohn-Sham (KS) approach is based on two theorems, known as Hohenberg-Kohn theorems [3].

First Theorem

The external potentials, which correspond to the nuclear-electron interaction potentials in the absence of an electromagnetic field, are determined by the electron density.

This implies that the ground-state properties of a molecule are completely determined by its electron density in the ground state, $\rho_0(x, y, z)$. This suggests that the ground-state energy (E_0) is a functional (function of a function) of ρ_0,

$$E_0 = E_0[\rho_0] \tag{9}$$

Second Theorem

The energy variational principle is always established for any electron density.

This suggests that any trial electron density (ρ_t) always leads to higher energy than the true ground-state energy, E_0. Note that the electronic energy obtained from a ρ_t is the energy of the electrons moving under the potential of the atomic nuclei, which is termed as an external potential (v) and therefore, the electronic energy is represented as $E_v = E_v[\rho_o]$

$$E_v[\rho_t] \geq E_0 \tag{10}$$

For a system of N electrons, ρ_t must follow the condition,

$$\int \rho_t(r)dr = N \tag{11}$$

Kohn-Sham Energy

Considering eq. (6), the ground-state energy (E_0) of any molecule is nothing but the kinetic energy plus potential energies due to the attraction between the nucleus (N) and electron (e) and the repulsion between two electrons. All of them are the functionals of ρ_0 and hence, the name density functional theory (DFT).

$$E_0 = <T[\rho_0]> + <V_{Ne}[\rho_0]> + <V_{ee}[\rho_0]> \tag{12}$$

$$<V_{Ne}> = \sum_{i=1}^{N} v(r_i) \tag{13}$$

$$<V_{Ne}[\rho_0]> = \int \rho_0(r) v(r) dr \tag{14}$$

Unfortunately, the functionals $T[\rho_0]$ and $V_{ee}[\rho_0]$ are not known. We consider a reference system of non-interacting electrons and define $\Delta<T[\rho_0]>$ and $\Delta<V_{ee}[\rho_0]>$ as the difference in the kinetic energy and electron-electron repulsion energy, respectively between the reference system and the actual (real) system:

$$\Delta<T[\rho_0]> = <T[\rho_0]> - <T_r[\rho_0]> \tag{15}$$

$$\Delta<V_{ee}[\rho_0]> = <V_{ee}[\rho_0]> - \frac{1}{2}\iint \frac{{}_0(_1){}_0(_1)}{r_{12}} dr_1 dr_2 \tag{16}$$

Substituting $<V_{Ne}[\rho_0]>$, $<T[\rho_0]>$ and $<V_{ee}[\rho_0]>$ from eq. (14), (15) and (16) into eq. (12) gives:

$$E_0 = \int \rho_0(r) v(r) dr + <T_r[\rho_0]> + \frac{1}{2}\iint \frac{\rho_0(r_1)\rho_0(r_2)}{r_{12}} dr_1 dr_2$$
$$+\Delta <T[\rho_0]> +\Delta <V_{ee}[\rho_0]> \tag{17}$$

Defining exchange-correlation energy functional, $E_{XC}[\rho_0]$

$$E_{XC}[\rho_0] = \Delta <T[\rho_0]> +\Delta <V_{ee}[\rho_0]> \tag{18}$$

$$E_0 = \int \rho_0(r) v(r) dr + <T_r[\rho_0]> + \frac{1}{2}\iint \frac{\rho_0(r_1)\rho_0(r_2)}{r_{12}} dr_1 dr_2 + E_{XC}[\rho_0] \tag{19}$$

The above eq. (19) contains four terms:

1st term:

$$1^{st} \text{ term: } \int \rho_0(r)v(r)dr = \int [\rho_0(r_i)\sum_A (-\frac{Z_A}{r_{iA}})] = -\sum_A Z_A \int \frac{\rho_0(r_i)}{r_{iA}}dr_i \qquad (20)$$

Once we obtain ρ_0, the integrals under the summation can be easily evaluated.

2nd term: The kinetic energy of the reference system with non-interacting electrons can be obtained as [4],

$$< T_r[\rho_0] > \ = \ <\psi_r \left| -\sum_{i=1}^{N} \nabla_i^2 \right| \psi_r > \qquad (21)$$

Since these electrons are non-interacting, ψ_r can be written as a single Slater determinant of occupied molecular orbitals. For a system of two electrons,

$$\psi_r = \frac{1}{\sqrt{2!}} \begin{vmatrix} \psi_1^{KS}(1)\alpha(1) & \psi_2^{KS}(1)\beta(1) \\ \psi_1^{KS}(2)\alpha(2) & \psi_2^{KS}(2)\beta(2) \end{vmatrix} \qquad (22)$$

The four components of the wavefunction in the determinant above represent the KS orbitals for the reference system. Each component appears as the product of a KS spatial orbital (ψ_i^{KS}) and spin function (α or β). Thus, eq. (21) can be easily solved for $< T_r[\rho_0]>$.

3rd term: The electronic repulsion energy can be easily evaluated once ρ_0 is obtained.

4th term: The only term we are left with is, exchange-correlation energy. DFT functional differs only in the way this term is incorporated!!

KS Equations and Solution

As per the second Hohenberg–Kohn theorem, the variational principle can be exploited to obtain the KS equations. For this purpose, we treat that the electron density of the reference system as the same as that of the actual system,

$$\rho_0 = \rho_r = \sum_{i=1}^{N} \left| \psi_i^{KS} \right|^2 \tag{23}$$

Substituting eq. (20), (21) and (23) back into eq. (19) and varying E_0 with respect to ψ_i^{KS} such that their orthonormality is preserved, we get the Kohn-Sham (KS) equations,

$$\left[-\frac{1}{2} \nabla_i^2 - \sum_A \frac{Z_A}{r_{1A}} + \int \frac{\rho(r_2)}{r_{12}} dr_2 + v_{XC}(1) \right] \psi_i^{KS}(1) = \varepsilon_i^{KS} \psi_i^{KS}(1) \tag{24}$$

Evidently, the KS equations are a set of equations for one-electron systems having terms ε_i^{KS} as the KS orbital energies (eigenvalues) and $v_{XC}(1)$ as the "exchange-correlation potential." The v_{XC} is obtained by taking the functional-derivative of $E_{XC}[\rho(r)]$ with respect to $\rho(r)$ as below,

$$v_{XC} = \frac{\delta E_{XC}[\rho]}{\delta \rho} \tag{25}$$

We can write the KS equations, eq. (24) in a compact form using the KS operator $\hat{h}^{KS}$ as below,

$$\hat{h}^{KS} \psi_i^{KS} = \varepsilon_i^{KS} \psi_i^{KS} \tag{26}$$

The KS equations, eq. (24) can be solved by expanding the KS orbitals, eq. (22), in terms of some basic functions φ_j,

$$\psi_i^{KS} = \sum_{j=1}^{M} c_{ij} \varphi_j \tag{27}$$

The eq. (27) can be substituted into the eq. (24) or (26) and then, the multiplication by $\varphi_1, \varphi_2 \ldots \varphi_m$ leads to the "M sets of M equations", that can be better represented as a matrix equation. This matrix is what is called the Fock matrix. The solution of the KS equations turns into the calculation of elements and diagonalization of the Fock matrix. The steps followed, subsequently, are as under:

1. Guess the density $\rho(r)$, usually by the summation of the electron densities of the individual atoms of the molecule, at the molecular geometry.

2. Obtain an explicit expression for the KS operator $\hat{h}^{KS}$,

3. Calculate the Fock matrix elements $h_{rs} = <\psi_r|\hat{h}^{KS}|\psi_s>$

4. Diagonalize the KS Fock matrix to obtain the coefficients c_{ij}.

5. Use these c_{ij} in eq. (27) to calculate better orbitals ψ_i^{KS} and hence, density function, eq. (23).

6. Use new density function to calculate better matrix elements, consequently, better c_{ij} which, in turn, provide a further improved density function.

7. Repeat this iterative process until the electron density converges.

8. Use final density and KS orbitals to calculate the energy.

THE EXCHANGE-CORRELATION ENERGY FUNCTIONAL

The exchange energy (E_X) is related to the exchange of two electrons of "same spin" whereas electron correlation energy (E_C) is related to the repulsion between two electrons of "different spins" occupying the same orbital. Both these effects lead to less overlapping of electron densities as compared to the reference system. The exchange-correlation energy functional (E_{XC}) can be written as the sum of an exchange-energy functional and a correlation-energy functional, both negative.

$$E_{XC} = E_X + E_C \tag{28}$$

As a matter of fact, E_X is much bigger than E_C in magnitude. For the argon atom, E_X is –30.19 Hartree, while E_C is only –0.72 Hartree [5].

The calculation of the E_{XC} and hence, v_{XC}, eq. (25) is very crucial as well as difficult. Since its inception, most of the research has been focused on this part of the theory. This is the part where we need some approximations:

The Local Density Approximation (LDA)

The LDA is the simplest form of approximation used for $E_{XC}[\rho]$. This is based on a homogeneous electron gas or the system with the electron density $\rho(r)$ varying only "slowly" with the position so that it could be considered as uniform. For every point, only the electron density "at that point" is considered, and hence, the term "local". In LDA, the exchange term is a simple analytical form obtained by using quantum Monte Carlo simulations [6] as below,

$$E_X^{LDA}[\rho] = -\frac{3}{4}\left(\frac{3}{\pi}\right)^{1/3}\int\rho^{4/3}(r)\,dr \tag{29}$$

Correlation functional is obtained at high density limit, *i.e.*, Wigner-Seitz radius, $r_s < 1$ [7] as below,

$$E_C^{LDA}[\rho] = C_1 + C_2\ln r_s + r_s\left(C_3 + C_4\ln r_s\right) \tag{30}$$

where C_1, C_2, C_3, and C_4 are some arbitrary constants. For the LDA, the E_{XC} and hence, v_{XC} can be accurately determined as follows,

$$E_{XC}^{LDA}[\rho] = \int\rho(r)\varepsilon_{XC}[\rho]dr \tag{31}$$

where ε_{XC} is the exchange-correlation energy per particle of homogenous electron gas.

Generalized Gradient Approximation (GGA)

Unlike homogenous electron gas, the electron density in an atom or molecule varies "greatly" and hence, the LDA has severe limitations. For instance, LDA overestimates the binding energy of the system. It does, therefore, not suffice to consider the electron density "locally" but requires some "non-local" methods. In non-local methods, both the electron density and its first derivatives with respect to position, *i.e.*, gradient are considered. The gradient of $\rho(r)$ at a point provides the sampling the value of ρ an infinitesimal distance beyond the "local" point of the coordinate r [8]. Such approximation is called the generalized-gradient approximation (GGA) and these functionals are known as the "gradient corrected" functional. However, it has been advised [9] not to use the term "non-local" while referring to the gradient-corrected functionals.

$$E_{XC}^{GGA}[\rho] = \int\rho(r)\varepsilon_{XC}^{GGA}\left[\rho(r),\nabla\rho(r)\right]dr \tag{32}$$

The general form of GGA in practice is expressed based on the LDA with an additional enhancement factor $F(s)$ that directly modifies the LDA energy:

$$E_{XC}^{GGA}[\rho,s] = \int \varepsilon_{XC}^{LDA}\left[\rho(r)]\rho(r)F(s)\right]dr \tag{33}$$

where

$$s = \frac{|\nabla\rho(r)|}{\rho^{4/3}(r)} \tag{34}$$

The typical value of s lies in the range 1-3.

The gradient corrections in the LDA have been proved to be more effective. The GGA suits almost all systems giving most structural properties within an error limit of 1-3% and corrects most of the over binding problems of the LDA. The practical DFT calculations developed after the introduction of the Becke 88 functional [10] for the exchange energy term. Examples of gradient-corrected correlation-energy functionals are the Lee–Yang–Parr (LYP) [11] and the Perdew and Wang 1991 (PW91) functional [12]. A newer GGA functional is PBE (Perdew-Burke-Ernzerhof, 1996) [13]. The PBE features the local electron density and its gradient, and second-order gradient in the enhancement factor.

Hybrid Functionals

These exchange-correlation energy functionals include a term of exchange energy calculated from the Hartree Fock (HF) method. As per the HF method, the electronic energy can be expressed as,

$$E = 2\sum_{i=1}^{N}H_{ii} + \sum_{i,j=1}^{N}(2J_{ij} - K_{ij}) \tag{35}$$

The first term, H_{ii} involves the kinetic energy of electron and electron–nucleus attraction and corresponding one-electron operator is defined as,

$$\hat{H}_i\varphi_i = \left[-\sum_{i=1}^{N}\frac{1}{2}\nabla_i^2 - \sum_{A=1}^{M}\sum_{i=1}^{N}\frac{Z_A}{r_{iA}}\right]\varphi_i$$

The second term, J_{ij} is the Coulomb integral corresponding to the Coulomb potential energy and it is defined by one-electron Coulomb operator as below,

$$\hat{J}_i\varphi_j = \varphi_j \int |\varphi_i|^2 \frac{1}{r_{ij}} dr_i$$

Here φ_i and φ_j are one-electron wavefunctions of ith and j^{th} electrons at a distance of r_{ij}. The third term, K_{ij} is associated with the one-electron exchange operator defined as,

$$\hat{K}_i\varphi_j(x_1) = \varphi_i(x_1) \int \frac{\varphi_i(x_2)\varphi_j(x_2)}{r_{12}} dx_2$$

Here $\varphi_i(x_1)$ and $\varphi_i(x_2)$ (and similarly φ_j's) are one-electron wavefunctions of ith electron as a function of positions of electron and their separation is r_{12}. The label 1 and 2 are just for convenience. The electrons are not distinguishable at all.

The exchange energy is related only to exchange integrals K_{ij} in the above equation. Therefore,

$$E_X = -\sum_{i=1}^{N}\sum_{j=1}^{N} K_{ij} \tag{36}$$

Substituting the KS orbitals into eq. (36), gives an expression for the HF exchange energy as below,

$$E_X^{\text{HF}} = -\sum_{i=1}^{N}\sum_{j=1}^{N} < \psi_i^{\text{KS}}(1)\psi_j^{\text{KS}}(2) \left| \frac{1}{r_{ij}} \right| \psi_i^{\text{KS}}(2)\psi_j^{\text{KS}}(1) > \tag{37}$$

Note that the KS Slater determinant describes the wavefunction of the reference system with non-interacting electrons "accurately" and hence, is the "exact" exchange energy for the reference system with the same electron density as that of the actual system. The accuracy of the molecular orbitals and consequently, exchange energy is, of course, affected by the basis set used.

A weighted contribution of the terms in the LDA and GGA functionals gives the expression for an HF/DFT exchange-correlation functional, generally referred to as a "hybrid" DFT functional. Such expressions are often parameterized to obtain the desired level of accuracy. So far, the most popular and "evergreen" hybrid

DFT functional is composed of an exchange-energy functional developed by Becke in 1993 and a correlation-energy functional devised by Lee, Yang, and Parr in 1988. This E_{XC} is termed as the Becke3LYP or B3LYP functional [10, 11], which is expressed as:

$$E_{XC}^{B3LYP} = \alpha E_X^{HF} + (1-\alpha)E_X^{HF} + \beta\Delta E_X^{Becke} + E_C^{local} + \gamma\Delta E_C^{LYP} \qquad (38)$$

Here first two terms incorporate the HF exchange energy functional from eq. (37), the third one is the exchange functional of Becke 88 as mentioned above, the fourth term is the "local" correlation part from the LDA and the last term is the LYP correlation functional described above. The parameters α, β, and γ are 0.20, 0.72, and 0.81, respectively. These values are chosen to give the best fit of the calculated value to molecular atomization energies.

In fact, the B3LYP is a gradient-corrected hybrid functional. It appears the most useful functional among many others which remain to be well-tested. This is the functional which will be employed in the succeeding chapters of this book. The DFT is, undoubtedly, waiting for further improved functionals, and the expectations from hybrid functional are very high as expressed by some pioneers [14]. There are, of course, several long-range corrected functionals such as CAM-B3LYP, M06, ωB97XD, *etc.* which improve certain results such as long-range interactions, excitation energies, hyperpolarizabilities and so forth.

Nevertheless, the speed of DFT calculations with gradient-corrected and/or hybrid functionals can be enhanced with a marginal loss in accuracy by the so-called perturbation method [15]. In this method, the KS eq. (24) are solved using the derivative eq. (25) using the LDA functional. Since LDA is simpler than the GGA or hybrid functional, the calculation is speeded up. Subsequently, the energy is obtained from eq. (19), now using the GGA or hybrid functional. This is how the things in computational research are managed. There are numerous functional available in the *Gaussian* program [16], see Chapter 2, a few of them are listed in Table **2** below:

Table 2. Some DFT functional available in the *Gaussian* program.

Functional	Type	Details
SVWN	LDA	S exchange and VWN correlation
BLYP	GGA	B exchange and LYP correlation
G96PW91		G96 exchange and PW91 correlation

(Table 2) cont.....

Functional	Type	Details
B3PW91	Hybrid	B exchange and PW91 correlation
B3LYP		B exchange and LYP correlation
PBE1PBE		PBE exchange and PBE correlation
CAM-B3LYP	Long-range corrected	Coulomb-attenuating method with B3LYP
M06		Minnesota exchange-correlation
ωB97XD		Dispersion corrected functional

THE BASIS SETS

A basis set is nothing but a mathematical description of the molecular orbitals (MOs). To describe the MOs accurately, a complete set of the basis functions are required. In principle, an infinite number of basis functions are needed for this purpose, but in practice, a finite number are employed [5]. The MOs are obtained by the linear combinations of a well-defined set of one-electron functions, known as the basis functions,

$$\varphi_i = \sum_{\mu=1}^{N} c_{\mu i} \chi_\mu \tag{39}$$

where $c_{\mu i}$ are referred to as the MO expansion coefficients and $\chi_1 \ldots \chi_N$ are the normalized basis functions. The *Gaussian* [16] (see Chapter 2) and several other programs utilize Gaussian-type functions to generate the basis sets. The (primitive) Gaussian functions can be expressed in the Cartesian form as below,

$$g(\alpha, r) = c x^n y^m z^l e^{-\alpha r^2} \tag{40}$$

where x , y , and z are components of and the constant α determines the size (radial extent) of the function. The normalization constant (c) is determined by,

$$\int_{\text{all space}} g^2 = 1 \tag{41}$$

These primitive Gaussians are employed as a linear combination in order to form the basis functions, which are known as the contracted Gaussians and expressed in the following form,

$$\chi_\mu = \sum_p d_{\mu p} g_p \tag{42}$$

where $d_{\mu p}$ are constants for a given basis set. Thus, the MOs can be represented for a basis set as follows,

$$\phi_i = \sum_\mu c_{\mu i} \chi_\mu = \sum_\mu c_{\mu i} \left(\sum_p d_{\mu p} g_p \right) \tag{43}$$

The accuracy of results in DFT calculation is largely determined by the size and quality of the basis set employed. So far, numerous basis sets have been optimized and tested. The quality of a basis set, *e.g.* 6-31G can be improved by increasing the number of basis functions per atom. The addition of polarization functions to the basis set, *e.g.* 6-31G* or 6-31G(d) includes the orbitals with angular momentum (*p, d, f, etc.*), which offers flexibility in different bonding situations. Likewise, the inclusion of diffuse functions (+), *e.g.* 6-31+G* or 6-31+G(d) is important for the systems with lone pairs, anions, and some excited states. The 6-31+G(d) is a valence double-zeta polarized basis set in which six primitive Gaussian functions are used for core atomic orbitals and valence orbitals are two basis functions comprising of three and one primitive Gaussians. In addition, five *d*-type Gaussian polarization and diffuse functions are included for each non-hydrogen atom.

The 6-311G** basis set, commonly used for electron correlation calculations on molecules containing first-row atoms, is valence triple-zeta, containing five *d*-type Gaussian polarization functions on each non-hydrogen atom and three *p*-type polarization functions on each hydrogen atom [17]. The size of a triple-zeta basis set [2] can be further increased by adding multiple polarization functions (3*d*, 3*df*, *etc.*) per atom or some extra functions for the valence orbitals. Apart from this, there are several correlation-consistent basis sets available, which are built up by adding shells of functions to a core set of basis functions. For instance, the cc-pVDZ (correlation consistent-polarized valence double zeta) basis set includes 1*s*, 1*p*, and 1*d* function. Likewise, the cc-pVTZ basis set adds another *s, p, d,* as well as an *f* functions, and so on. The augmentations to these basis sets have been made by adding diffuse functions to better describe anions and weakly interacting molecules, such as aug-cc-pVDZ, aug-cc-pVTZ, aug-cc-pVQZ, *etc.* The common (split-valence) basis sets available in the *Gaussian* (as well as other programs) are listed in Table **3**. More basis sets can be obtained from the basis set exchange library [18].

Table 3. The basis sets available in the *Gaussian* program.

Basis Set	Quality	Number of Basis Functions for Valence Orbitals
6-31G	double-zeta	2 comprising of 3 and 1 primitive Gaussians
6-31G*	double-zeta	add 5 d-type polarization functions for non H atoms
6-31G**	double-zeta	add 3 p-type polarization functions for H atoms
6-31+G**	double-zeta	add diffuse function for non H atoms
6-31++G**	double-zeta	add diffuse function for H atoms
6-311G	triple-zeta	3 comprising of 3, 1 and 1 primitive Gaussians
6-311++G**	triple-zeta	add 5 d-type polarization and diffuse functions for non H atoms and 3 p-type polarization and diffuse functions for H atoms

PROS AND CONS OF DFT

Pros

DFT incorporates electron correlation effects in its theoretical basis, unlike the wavefunction-based methods such as MP2, CC, CI, *etc.* in which the correlation effects are to be included explicitly in the framework of the HF theory. Therefore, the DFT offers the accuracy comparable to MP2 calculations, but at approximately the same computational cost (time) as needed for the HF calculations. This makes DFT a cost-effective approach.

Unlike wavefunction based methods, DFT becomes basis-set saturated more easily. Therefore, the DFT calculations are feasible on larger systems than the wavefunction-based methods can handle. Note that the DFT is considered to be a universal choice for transition metal compounds, as the wavefunction-based methods perform very poorly [19].

The DFT is based on the electron density, an observable unlike wavefunction, a mathematical entity. The results obtained from the DFT can be easily and intuitively grasped [20].

Cons

The core of DFT, which is the exchange-correlation functional, is not exactly known and there is no systematic way to improve our approximations to it. Most of the functionals, but not all, are generally modified based on experience as well as intuition and testing the calculations against experiments (benchmarking). However, some of them can also be constructed based on physical conditions and corrections. In the wavefunction-based methods, on the contrary, bigger basis sets and higher correlation levels are likely to turn towards an exact solution of the Schrödinger equation.

Furthermore, these functionals are not derived from a theoretical framework and therefore, the application of DFT on novel molecular systems becomes sometimes questionable. Moreover, only the approximate nature of functional destroys the variational form of DFT, *i.e.*, the DFT calculated energy might be lower than the actual energy value.

The accuracy of DFT is often limited when compared with the highest-level wavefunction-based methods, such as QCISD(T) and CCSD(T). Even the most popular hybrid functional such as B3LYP is not able to deal with the van der Waals interactions [21, 22]. Some GGA functionals as well as long-range corrected functionals, of course, offer reasonable structures and energies of hydrogen-bonded species [23, 24].

Unlike wavefunction-based methods, DFT is of single-configurational character. This limits DFT to provide the bond dissociation potential energy surface. This is probably the most serious disadvantage of DFT.

Basically, the DFT is a ground-state theory. However, several methods have been developed for extending it to excited states. One of the methods is described below. Note that the development is still continued.

TIME-DEPENDENT DENSITY FUNCTIONAL THEORY (TDDFT)

The TDDFT extends the concept of DFT to time-dependent situations. For time-dependent situations, eq. (1) reads

$$i\frac{\partial}{\partial t}\Psi(t) = \hat{H}(t)\Psi(t) \tag{44}$$

where

$$\hat{H}(t) = -\sum_{i=1}^{N}\frac{1}{2}\nabla_i^2 + \sum_{i=1}^{N}\sum_{j>i}^{N}\frac{1}{r_{ij}} - \sum_{i=1}^{N}v_{ext}(r_i,t) \tag{45}$$

For a multi-electron system, there is a one-to-one mapping between electron density and external potential [25]. Analogous to the time-independent situations, all observables of any system, in the case of a time-dependent potential, can be uniquely determined by a time-dependent density $\rho(r,t)$ and its state at an arbitrary (single) instant of time $\psi(0)$. This is known as the Runge-Gross theorem.

If the system has been in its ground state until the time-dependent potential is switched on at any time t_0, all observables are functionals of the density only. To put it another way, the initial state of the system $\psi(0)$ at time t_0 is a unique functional of the density at t_0. This density is similar to the ground-state density of the stationary system before t_0.

This "unique" relationship leads to the development of a computational framework in which the effect of the interaction between particles corresponds to a density-dependent single-particle potential. This enables the study of the time evolution of an interacting system by treating a time-dependent auxiliary single-particle problem. The TD KS equations can be written as:

$$i\frac{\partial \psi_j(r,t)}{\partial t} = \left[-\frac{\nabla_j^2}{2} + v_j^{KS}(r,t) \right] \psi_j(r,t) \tag{46}$$

such that

$$\rho(r,t) = \sum_{i=1}^{N} \left| \psi_j(r,t) \right|^2 \tag{47}$$

The time-dependent KS potential is defined as:

$$v^{KS}(r,t) = v_{ext}(r,t) + v_H(r,t) + v_{XC}(r,t) \tag{48}$$

The first term is analogous to that defined by eq. (20). The second term classical repulsion energy as earlier. The third term is the exchange-correlation term. It is a functional of time-dependent electron density, $\rho(r,t)$ and initial state, $\psi(0)$. Thus, the knowledge of $v_{XC}(r,t)$ gives the solution to time-dependent problems. Here again, the major issue of TDDFT is to obtain appropriate and reasonable approximations for the exchange-correlation component of the time-dependent KS potential. Initially, Runge and Gross defined this KS potential based on the Dirac action [25], however, it was not in accordance with the causality of the (density) response functions [26], *i.e.*, the functional derivative of the density with respect to the external potential. For instance, a change in the potential at a time $t = t_0$ can not affect the density at time $t < t_0$. Later, this problem was removed by an alternative formalism [27].

Usually TDDFT is based on Casida's linear response theory [28]. According to this theory, if the external potential (perturbation) is small such that it does not completely destroy the ground-state density, one can expect the linear response of the system. This implies that the variation in the system, to the first approximation, will depend only on the ground-state density so that we can simply use all the properties of DFT. The readers are referred to [28] for a detailed discussion on linear response TDDFT.

I will finish this chapter with a quote from a famous quantum physicist P. A. M. Dirac:

"The fundamental laws necessary for the mathematical treatment of a large part of physics and the whole of chemistry are thus completely known, and the difficulty lies only in the fact that application of these laws leads to equations that are too complex to be solved."

CONCLUDING REMARKS

In this chapter, we have discussed the formulation of the DFT and TDDFT. The important aspect of (TD)DFT is the approximation used for exchange-correlation functional. There are numerous functionals available under various approximations, be it LDA or GGA, or hybrid. Therefore, the choice of an appropriate functional for the system under study becomes very crucial. Along with a suitable basis set, an appropriate DFT method is capable to reproduce the experimental data. In this regard, a hybrid B3LYP functional becomes the first choice with a double- or triple-zeta basis set. In most of the succeeding chapters in this book, or perhaps all, the B3LYP method will be employed. Nevertheless, the choice of functional largely depends on the nature of the system, property under study, desired level of accuracy, and available computational resources.

CONSENT FOR PUBLICATION

Not applicable.

CONFLICT OF INTEREST

The authors declare no conflict of interest, financial or otherwise.

ACKNOWLEDGEMENTS

The author acknowledges the funding from University Grant Commission, India through Startup grant [30-466/2019(BSR)].

REFERENCES

[1]   Pople, J.A.; Beveridge, D.L. *Approximate Molecular Orbital Theory*; McGraw-Hill: New York, **1970**.

[2]   Foresman, J.B.; Frisch, A. *Exploring Chemistry with Electronic Structure Methods*; Gaussian Inc: Pittsburgh, **1993**.

[3]   Hohenberg, P. P.; W. Kohn, W. Density functional theory (DFT). *Phys. Rev. B,* **1964**, *136*, 864.
[http://dx.doi.org/10.1103/PhysRev.136.B864]

[4]   Parr, R.G.; Yang, W. *Density-Functional Theory of Atoms and Molecules*; Oxford, New York, **1989**.

[5]   Levine, I.N. *Quantum Chemistry*; Prentice Hall: Upper Saddle River, New Jersey, **2000**.

[6]   Ceperley, D.M.; Alder, B.J. Ground state of the electron gas by a stochastic method. *Phys. Rev. Lett.,* **1980**, *45*, 566.
[http://dx.doi.org/10.1103/PhysRevLett.45.566]

[7]   Perdew, J.P.; Zunger, A. Self-interaction correction to density-functional approximations for many-electron systems. *Phys. Rev. B Condens. Matter,* **1981**, *23*, 5048.
[http://dx.doi.org/10.1103/PhysRevB.23.5048]

[8]   Apostol, T.M. *Mathematical Analysis*; Addison-Wesley: Reading, MA, **1957**.

[9]   St. Amant, A. *Reviews in Computational Chemistry*; Wiley VCH: New York, **1996**, 7, .

[10]   Becke, A.D. Density-functional exchange-energy approximation with correct asymptotic behavior. *Phys. Rev. A Gen. Phys.,* **1988**, *38*(6), 3098-3100.
[http://dx.doi.org/10.1103/PhysRevA.38.3098] [PMID: 9900728]

[11]   Lee, C.; Yang, W.; Parr, R.G. Development of the Colle-Salvetti correlation-energy formula into a functional of the electron density. *Phys. Rev. B Condens. Matter,* **1988**, *37*(2), 785-789.
[http://dx.doi.org/10.1103/PhysRevB.37.785] [PMID: 9944570]

[12]   Perdew, J.P.; Wang, Y. Pair-distribution function and its coupling-constant average for the spin-polarized electron gas. *Phys. Rev. B Condens. Matter,* **1992**, *46*(20), 12947-12954.
[http://dx.doi.org/10.1103/PhysRevB.46.12947] [PMID: 10003333]

[13]   Perdew, J.P.; Burke, K.; Wang, Y. Generalized gradient approximation for the exchange-correlation hole of a many-electron system. *Phys. Rev. B Condens. Matter,* **1996**, *54*(23), 16533-16539.
[http://dx.doi.org/10.1103/PhysRevB.54.16533] [PMID: 9985776]

[14]   Kohn, W.; Becke, A.D.; Parr, R.G. Density functional theory of electronic structure. *J. Phys. Chem.,* **1996**, *100*, 12974-12980.
[http://dx.doi.org/10.1021/jp960669l]

[15]   Fan, L.; Ziegler, T. The influence of self-consistency on nonlocal density functional calculations. *J. Chem. Phys.,* **1991**, *94*, 6057-6063.
[http://dx.doi.org/10.1063/1.460444]

[16]   Frisch, M.J.; Trucks, G.W.; Schlegel, H.B. *Gaussian 09, Revision B.01*; Gaussian Inc: Wallingford, CT, **2010**.

[17]   Čížek, J. On the correlation problem in atomic and molecular systems. Calculation of wavefunction components in Ursell-type expansion using quantum-field theoretical methods. *J. Chem. Phys.,* **1966**, *45*, 4256-4266.
[http://dx.doi.org/10.1063/1.1727484]

[18]   https://www.basissetexchange.org/

[19]   Bauschlicher, C.W., Jr; Ricca, A.; Partridge, H.; Langhoff, S.R. *Recent Advances in Density Functional Methods. Part II*; World Scientific: Singapore, **1995**.

[20]   Shusterman, A.J.; Shusterman, G.P. Teaching chemistry with electron density models. *J. Chem. Educ.,* **1997**, *74*, 771.

[http://dx.doi.org/10.1021/ed074p771]

[21]   Kristyán, S.; Pulay, P. Can (semi) local density functional theory account for the London dispersion forces? *Chem. Phys. Lett.,* **1994**, *229*, 175-180.
[http://dx.doi.org/10.1016/0009-2614(94)01027-7]

[22]   Pérez-Jordá, J.; Becke, A.D. A density-functional study of van der Waals forces: rare gas diatomics. *Chem. Phys. Lett.,* **1995**, *233*, 134-137.
[http://dx.doi.org/10.1016/0009-2614(94)01402-H]

[23]   Lozynski, M.; Rusinska-Roszak, D.; Mack, H.G. Hydrogen bonding and density functional calculations: The B3LYP approach as the shortest way to MP2 results. *J. Phys. Chem. A,* **1998**, *102*, 2899-2903.
[http://dx.doi.org/10.1021/jp973142x]

[24]   Sim, F.; St. Amant, A.; Papai, I.; Salahub, D.R. Gaussian density functional calculations on hydrogen-bonded systems. *J. Am. Chem. Soc.,* **1992**, *114*, 4391-4400.
[http://dx.doi.org/10.1021/ja00037a055]

[25]   Runge, E.; Gross, E.K. Density-functional theory for time-dependent systems. *Phys. Rev. Lett.,* **1984**, *52*, 997.
[http://dx.doi.org/10.1103/PhysRevLett.52.997]

[26]   van Leeuwen, R. Causality and symmetry in time-dependent density-functional theory. *Phys. Rev. Lett.,* **1998**, *80*, 1280-1283.
[http://dx.doi.org/10.1103/PhysRevLett.80.1280]

[27]   Vignale, G. Real-time resolution of the causality paradox of time-dependent density-functional theory. *Phys. Rev. A,* **2008**, *77*, 062511.
[http://dx.doi.org/10.1103/PhysRevA.77.062511]

[28]   Casida, M.E.; Jamorski, C.; Bohr, F.; Guan, J.; Salahub, D.R. *Theoretical and Computational Modeling of NLO and Electronic Materials. Karna, S. P*; Yeates, A.T., Ed.; ACS Press: Washington, D.C, **1996**.

CHAPTER 2

Applications of DFT on Molecular Systems: How Gaussian Works

Abstract: This chapter focuses on the practical application of DFT in molecular systems. We discuss the process of "geometry optimization" and the idea behind it, which is the very first step of every DFT calculation. We introduce the *Gaussian*, a popular software program to perform such calculations. We continue to discuss the capability of this program with a brief theoretical background, wherever needed. We talk about several kinds of calculations to be performed by the *Gaussian* such as thermodynamics, population analysis, NMR, NLO, NBO, TDDFT calculations to name a few. More importantly, we discuss how to perform these calculations, extract, and interpret the results. Ideally, this chapter provides all the ingredients needed to grasp the results discussed in the forthcoming chapters.

Keywords: DFT, Frequency calculation, Gaussian, GaussView, Geometry optimization, MESP, Molecular orbital, NBO, NLO, NMR, Population analysis, TDDFT, Thermodynamics, UV-vis-NIR.

INTRODUCTION

Having chosen an appropriate DFT method, *i.e.*, the suitable combination of exchange-correlation functional and basis set, the primary task of any DFT calculation is to obtain the "correct" structure of a molecular system. This is what is known as the geometry optimization of a molecule. The initial configuration of the molecular system required for this task can be derived from crystallographic parameters or modelled hypothetically. The process of geometry optimization has to be followed by the vibrational frequency calculation. This is to ensure that the optimized configuration of a molecule is stable, *i.e.*, its energy is "truly" minimum. Therefore, we shall first discuss these two terms.

Further applications of DFT on the molecular systems depend on the nature and/or property of the systems under study. For instance, if the molecule contains an electron-donating or withdrawing group, one might be interested in the charge distribution. Likewise, if the molecule is asymmetric, one might be interested in its optical activity and so forth.

Ambrish Kumar Srivastava and Neeraj Misra

GEOMETRY OPTIMIZATION AND FREQUENCY CALCULATIONS

The objective of geometry optimization is to find an atomic arrangement that makes the molecule most stable. Molecules are most stable when their energy attains the lowest possible value. This is achieved by creating a "potential energy surface" (PES). Before we discuss the PES, let's first have a look over the minimization of a potential energy function in one-dimension, $V(x)$.

A "stationary point" of $V(x)$ is characterized by:

$$\frac{dV(x)}{dx} = 0 \tag{1}$$

The $V(x)$ is said to have a minimum if,

$$\frac{d^2V(x)}{dx^2} > 0 \tag{2}$$

and it has a maximum if,

$$\frac{d^2V(x)}{dx^2} < 0 \tag{3}$$

But it is neither maximum nor minimum for,

$$\frac{d^2V(x)}{dx^2} = 0 \tag{4}$$

In Fig. (**1**), we have plotted the $V(x)$, which shows the minimum and maximum stationary points. The "global" minimum is the point that is the lowest among all local minima.

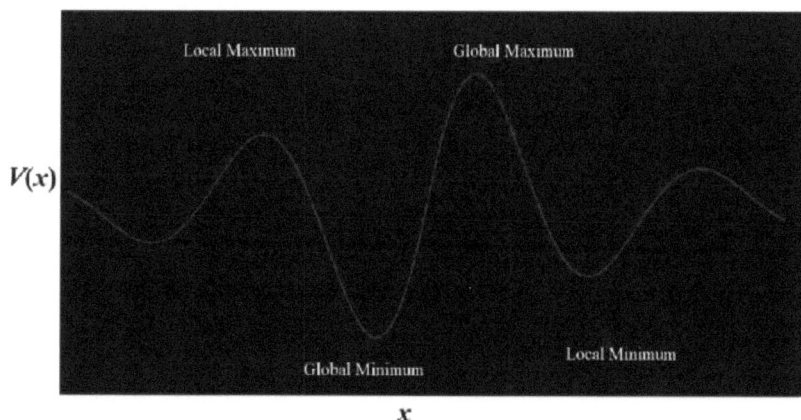

Fig. (1). Variation of *V(x)* with respect to the coordinate x. Various stationary points have also been shown.

The generalization of $V(x)$ to the three dimensions leads to the concept of a PES. Thus, the PES is a mathematical relationship between different molecular geometries and their corresponding energies. These values are usually displayed in a three-dimensional graph, representing the bond angle, bond distance and, energy values as shown in Fig. (**2**).

For a molecular system of N atoms, there are $3N$ coordinates (r). Therefore, eq. (1) can be written as follows:

$$|\nabla V(\vec{r}^{3N})| = 0 \tag{5}$$

The energy of a molecule system near its equilibrium structure (stationary points) becomes:

$$E = T + V = \frac{1}{2}\sum_{i=1}^{3N} q_i^2 + V_{eq} + \sum_{i=1}^{3N}\sum_{j=1}^{3N}\left(\frac{\partial^2 V}{\partial q_i \partial q_j}\right)_{eq} q_i q_j \tag{6}$$

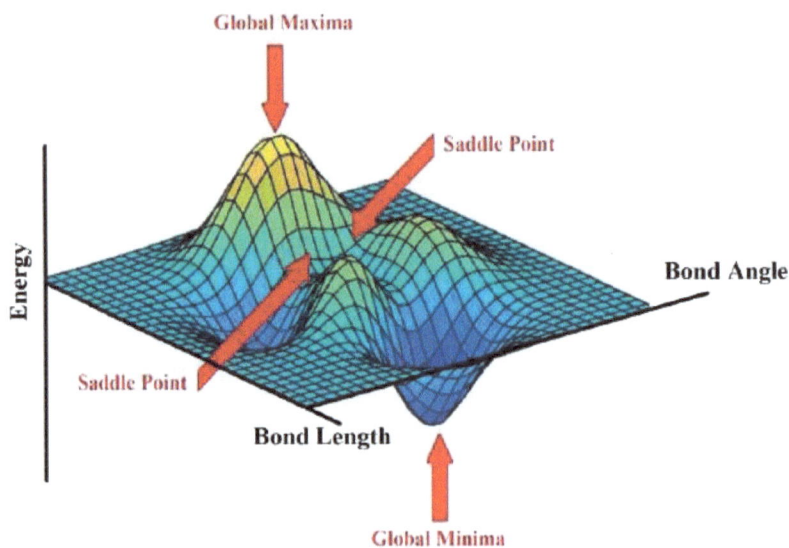

Fig. (2). Variation of *V(x)* with respect to the coordinate x. Various stationary points have also been shown.

Here the mass-weighted Cartesian displacements, q_i, are defined in terms of the locations X_i of the nuclei relative to their equilibrium positions $X_{i'\text{eq}}$ and their masses M_i,

$$q_i = M_i^{1/2} \left(X_i - X_{i'\text{eq}} \right) \tag{7}$$

V_{eq} is the potential energy at the equilibrium nuclear configuration.

The power series expansion in eq. (6) is "truncated" at second order [1] and therefore, the equation of motion takes the form,

$$q_j = \sum_{i=1}^{3N} H_{ij} q_i \; ; \; j = 1, 2, 3 \dots 3N \tag{8}$$

The components H_{ij} are quadratic force constants, which can be expressed as the second derivatives of the potential energy with respect to mass-weighted Cartesian displacement, evaluated at the equilibrium nuclear configuration, *i.e.*,

$$H_{ij} = \left(\frac{\partial^2 V}{\partial q_i \partial q_j} \right)_{\text{eq}} \tag{9}$$

The H_{ij} may be evaluated by numerical differentiation, either by taking numerical first differentiation of analytical first derivatives or by direct analytical second differentiation. In the first case, we have:

$$\frac{\partial^2 V}{\partial q_i \partial q_j} = \frac{\Delta\left(\partial V / \partial q_j\right)}{\Delta q_i} \tag{10}$$

Once we obtain the H_{ij} components, we can immediately recognize $[H_{ij}]$ as a real and symmetric $3N3N$ matrix, which is known as the "Hessian" having eigenvalues (λ_H).

The condition of local minimum in case of molecular systems becomes,

$$\{\lambda_H\} \geq 0 \tag{11}$$

The eigenvalues of Hessian provide the normal-mode vibrational frequencies. For N-atomic systems, eq. (6) can be solved by standard procedures [2] to yield a set of $3N$ normal modes. However, six of these (five for linear molecules) will be zero as they do not correspond to vibrational but translational and rotational degrees of freedom. Therefore, the number of normal modes for molecular systems becomes,

$$k = 3N - 6 \tag{12}$$

The PES is characterized by various distinct points (see Fig. **2**) as listed in Table **1** below:

Table 1. Various stationary points on the PES of any molecular system.

Stationary Points	Interpretation in Terms of Energy
Local maximum	The highest value in a particular section or region of the PES
Local minimum	The lowest value in a particular section or region of the PES
Global maximum	The highest value in the entire PES
Global minimum	The lowest value in the entire PES
Saddle point	Maximum in one direction and a minimum in the other.

Note that a global minimum represents the most optimal molecular geometry. The PES, therefore, gives us a quick way to look and approximate a reasonable geometry.

In terms of λ_H, a global/local maximum is characterized by:

$$\{\lambda_H\} < 0 \tag{13}$$

In this case, the structure of a molecule cannot be optimized by having several negative (imaginary) frequencies, *i.e.*, it is not energetically stable. Likewise, a saddle point is characterized by "only one" negative value of frequency. Saddle points correspond to a transition state connecting two equilibrium structures, which are very important in the study of reaction kinetics and mechanisms.

Frequency calculation is not only required to obtain the equilibrium (optimized) structure of molecules but also yields some of the most important aspects covered in this book. Vibrational infrared (and also Raman) spectra of molecules can be predicted for any optimized molecular structure. The position and relative intensity of vibrational bands can be gathered from the results of a frequency calculation. This information is independent of the experiment and can therefore be used as a tool to confirm peak positions in experimental FT-IR and FT-Raman spectra or to predict peak positions and intensities when experimental data is not available.

These frequencies are "generally" calculated on the basis of the harmonic approximation, while anharmonicity is always present in actual vibrations. This partially explains the discrepancies between calculated and experimental frequencies. That's why the calculated frequencies are generally scaled by some factor (depending on the method of computation) prior to comparison with experimental data obtained from FT-IR, FT-Raman, *etc.* The scaling of vibrational frequencies will be discussed in detail in chapter 3.

INTRODUCTION TO THE *GAUSSIAN*

The DFT is implemented in almost all software programs, be it free or commercial. The freely available software programs include *GAMESS* [3], *Firefly* [4], *ORCA* [5], *etc.* The *Gaussian* [6] is the most popular "commercial" software package. The results presented in this book are obtained from the *Gaussian* and therefore, it is relevant to introduce the *Gaussian* and its capabilities.

The *Gaussian* is currently distributed by Gaussian Inc. Wallingford, USA [7]. The current version of the program is *Gaussian 16*. The *GaussView* is in-house modelling and visualization software, which is distributed along with the *Gaussian* in its latest version, *GaussView* 6. *GaussView* is used to model the molecular system, create the input (.gjf file) and visualize the output (.log or .out file). The *Gaussian* processes the input created by *GaussView* and perform all calculations, and produces an output. The graphical user interfaces (GUIs) of *Gaussian* and *GaussView* programs are displayed in Figs. (**3** and **4**), respectively.

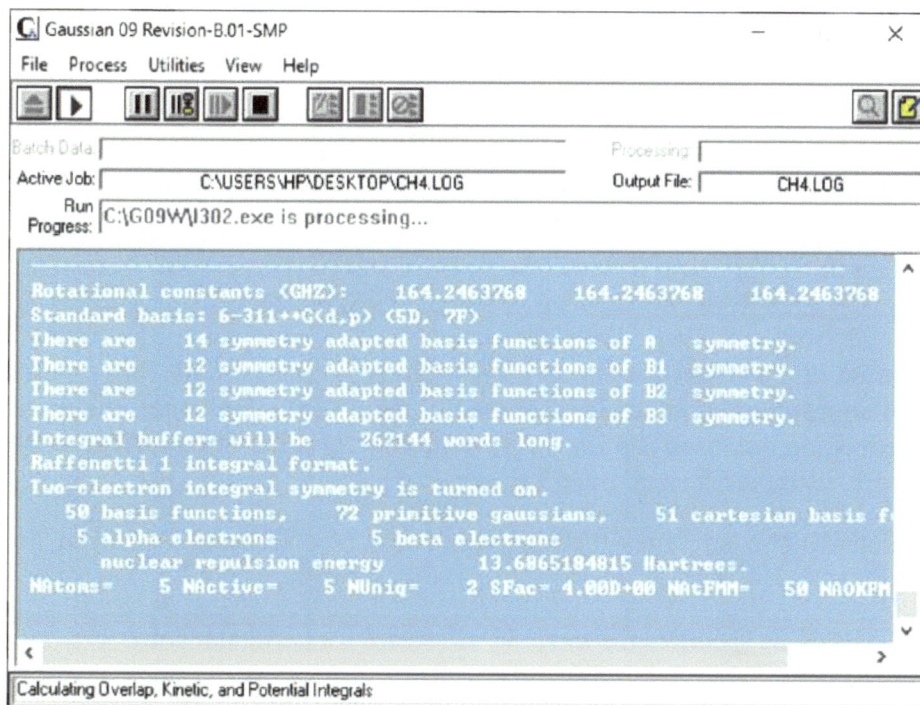

Fig. (3). Graphical User Interface of the *Gaussian*.

The input for the *Gaussian* can be created in several ways. One can use the crystallographic parameters and convert them to the Cartesian coordinates, which can be used in the *Gaussian* (Fig. **3**). However, the easiest and more intuitive way is to model the structure with the help of *GaussView* (Fig. **4**). There are mainly two panels of the *GaussView*: Sketcher and Visualizer. The Sketcher (the uppermost panel in Fig. (**4**) comes with plenty of options/menus. The Visualizer (the rightmost small panel) is used to display the structure and visualize other properties. The structure is created in Visualizer with the help/options of Sketcher.

Once the structure is modelled, one has to go to the *Calculate* menu in the Sketcher panel and click on the *Gaussian Calculation Setup*. This opens an additional panel, shown as the leftmost panel in Fig. (**4**). Now, one needs to select the *Job Type*. There are 8 basic jobs enlisted under the *Job Type* option, which are elaborated in Table **2**. Of course, there are numerous other jobs available in the *Gaussian*. A complete list can be obtained from the *Gaussian* website [8].

Table 2. Types of basic jobs to be performed in the *Gaussian*.

Job Type	Elaboration
Energy	To calculate single-point energy for a given structure
Optimization	To optimize the given structure and obtain a stationary point
Frequency	To calculate the vibrational frequency for a given structure
Opt+Freq	To perform Optimization followed by Frequency calculations
IRC	To integrate intrinsic reaction coordinate during a reaction path
Scan	To scan a potential energy surface of a given structure
Stability	To test the stability of wavefunction during DFT calculations
NMR	To predict NMR shielding tensors and magnetic susceptibilities

The next option in the *Gaussian Calculation Setup* is to choose the *Method*. After choosing an appropriate method, as discussed in the preceding chapter, the input can be saved as the *gaussian job file* (.gjf). Alternatively, the input (.gjf) can be created directly using the format given below:

%nprocshared=4 (This specifies the number of processors/cores to be used for Job)

%mem=500MB (This specifies the memory allotted for a particular Job)

%chk=F:\ch4.chk (This specifies the path for checkpoint file)

opt freq b3lyp/6-311+g(d) geom=connectivity (Method and Keyword)

Methane Job (Job Title)

0 1 (Spin and Multiplicity)

C 0.15037595 1.84210524 0.00000000 (Atom and Input Cartesian coordiantes)

H 0.50703037 0.83329523 0.00000000

H 0.50704879 2.34650343 0.87365150

H 0.50704879 2.34650343 -0.87365150

H -0.91962405 1.84211842 0.00000000

1 2 1.0 3 1.0 4 1.0 5 1.0 (Connectivity of Atoms)

2

3

4

5

Subsequently, it can be run by the *Gaussian*. The processing of the job for methane (CH_4) molecule can be seen in Fig. (**3**).

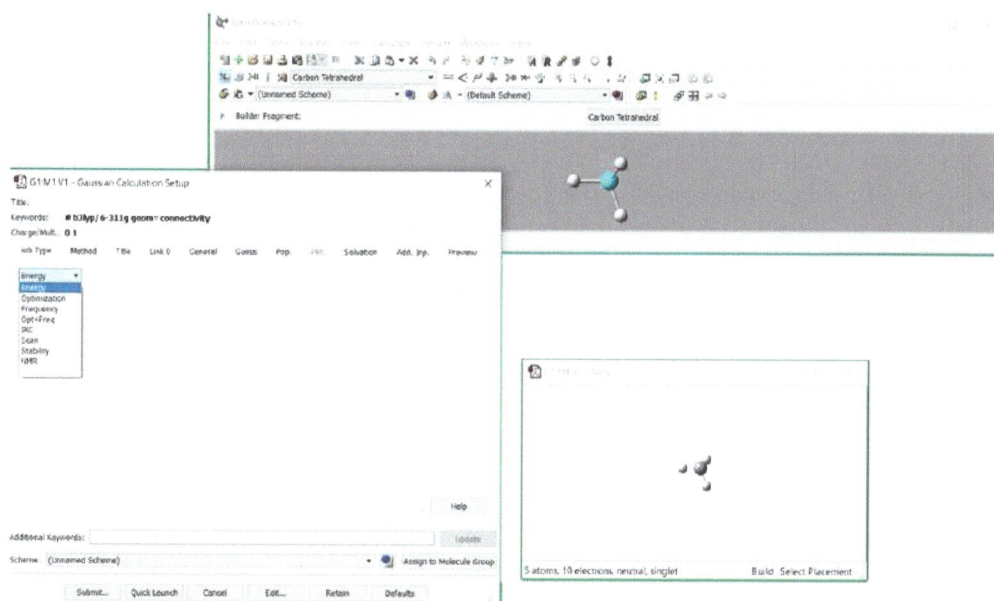

Fig. (4). Graphical User Interface of the *GaussView*.

Once the job is finished, you are likely to obtain an output (.log) file and a checkpoint (.chk) file. The output file contains all the results of the calculation, except grid data required for plotting various surfaces. The latter can be obtained in the checkpoint file. The output file for CH_4 is given in Appendix for the "Opt+Freq" job. Below we interpret the results of this calculation using the snapshots of the output file.

----

opt freq b3lyp/6-311+g(d) geom=connectivity

1/14=-1,18=20,19=15,26=3,38=1,57=2/1,3;

2/9=110,12=2,17=6,18=5,40=1/2;

3/5=4,6=6,7=11,11=2,16=1,25=1,30=1,71=1,74=-5/1,2,3;

4//1;

5/5=2,38=5/2;

The first line is simply the job requested followed by the method used. We have chosen the "Opt+Freq" job and the B3LYP/6-311+G(d) method. The numbers below are default internal options (IOp) of the Gaussian for the requested job and the method used. More about IOp can be found here [9].

Methane Job

Symbolic Z-matrix:

Charge = 0 Multiplicity = 1

C 1.76692 0.56391 0.

H 2.12357 -0.4449 0.

H 2.12359 1.06831 0.87365

H 2.12359 1.06831 -0.87365

H 0.69692 0.56392 0.

This is the model structure of CH_4 in Z-matrix format, which is used to obtain initial parameters such as bond lengths, bond angles, *etc.* of the molecule.

! Initial Parameters !

! (Angstroms and Degrees) !

-------------------------- --------------------------

! Name Definition Value Derivative Info. !

! R1 R(1,2) 1.07 estimate D2E/DX2 !

! R2 R(1,3) 1.07 estimate D2E/DX2 !

! R3 R(1,4) 1.07 estimate D2E/DX2 !

! R4 R(1,5) 1.07 estimate D2E/DX2 !

! A1 A(2,1,3) 109.4712 estimate D2E/DX2 !

Below this appears the input orientation in Cartesian as well as distance matrix format. The job processing starts, hereafter, the self-consistent field (SCF) iterations.

Item Value Threshold Converged?

Maximum Force 0.000152 0.000450 YES

RMS Force 0.000081 0.000300 YES

Maximum Displacement 0.000434 0.001800 YES

RMS Displacement 0.000232 0.001200 YES

Predicted change in Energy=-1.239284D-07

Optimization completed.

-- Stationary point found.

This confirms that a stationary point is found, *i.e.*, the optimization of a molecule is completed.

```
-----------------------------
```

! Optimized Parameters !

! (Angstroms and Degrees) !

```
------------------------- -------------------------
```

! Name Definition Value Derivative Info. !

```
---------------------------------------------------
```

! R1 R(1,2) 1.0905 -DE/DX = 0.0002 !

! R2 R(1,3) 1.0905 -DE/DX = 0.0002 !

! R3 R(1,4) 1.0905 -DE/DX = 0.0002 !

! R4 R(1,5) 1.0905 -DE/DX = 0.0002 !

This provides the optimized parameters of the molecule. Note that only the first derivatives of the energy with respect to the molecular coordinate have been evaluated at this stage, eq. (5). Evidently, it does not guarantee for the stationary point to be a minimum. This demands the calculation of frequencies and therefore, we have requested the "Opt+Freq" job.

Link1: Proceeding to internal job step number 2.

```
------------------------------------------------------------------
```

#N Geom=AllCheck Guess=TCheck SCRF=Check GenChk RB3LYP/6-311+G(d) Freq

```
------------------------------------------------------------------
```

1/10=4,29=7,30=1,38=1,40=1/1,3;

2/12=2,40=1/2;

3/5=4,6=6,7=11,11=2,14=-4,16=1,25=1,30=1,70=2,71=2,74=-5,116=1/1,2,3;

4/5=101/1;

5/5=2,98=1/2;

Now, the job will proceed to the frequency calculations.

! Initial Parameters !

! (Angstroms and Degrees) !

------------------------- -------------------------

! Name Definition Value Derivative Info. !

! R1 R(1,2) 1.0905 calculate D2E/DX2 analytically !

! R2 R(1,3) 1.0905 calculate D2E/DX2 analytically !

! R3 R(1,4) 1.0905 calculate D2E/DX2 analytically !

! R4 R(1,5) 1.0905 calculate D2E/DX2 analytically !

! A1 A(2,1,3) 109.4712 calculate D2E/DX2 analytically !

These are the initial parameters, which are to be used for frequency calculations. Unlike the initial parameters used for the optimization, these are "actually" optimized parameters. One must optimize the structure of a molecule before calculating frequencies.

Full mass-weighted force constant matrix:

Low frequencies --- -0.0012 -0.0011 -0.0003 46.1637 46.1637 46.1637

Low frequencies --- 1355.1285 1355.1285 1355.1285

Harmonic frequencies (cm**-1), IR intensities (KM/Mole), Raman scattering

activities (A**4/AMU), depolarization ratios for plane and unpolarized

incident light, reduced masses (AMU), force constants (mDyne/A), and normal coordinates:

1 2 3

T2 T2T2

Frequencies -- 1355.1285 1355.1285 1355.1285

This is the portion where the results of frequency calculations appear, as per eq. (9).

1|1|UNPC-LAPTOP-VBAQNH7H|Freq|RB3LYP|6-311+G(d)|C1H4|HP|29-Nov-2020|0|

|#N Geom=AllCheck Guess=TCheck SCRF=Check GenChk RB3LYP/6-311+G(d) Fre

q||Title Card Required||0,1|C,1.76691739,0.563909725,0.|H,2.1304079871

,-0.4642365757,-0.0000009202|H,2.1304274232,1.077975738,0.8903970013|H

,2.1304259205,1.0779768006,-0.8903970013|H,0.6764082293,0.5639229371,0

.0000009202||Version=IA32W-G09RevB.01|State=1-A1|H-=-40.5280819|RMSD=7

.046e-010|RMSF=7.840e-005|ZeroPoint=0.0448027|Thermal=0.047669|Dipole=

0.,0.,0.|DipoleDeriv=-0.015942,0.,0.,0.,-0.015942,0.,0.,0.,-0.015942,0

.0578811,0.076215,0.,0.076215,-0.1307508,-0.0000002,0.,-0.0000002,0.08

48261,0.0578782,-0.038109,-0.0660074,-0.038109,0.0309334,-0.0933458,-0

.0660074,-0.0933458,-0.0768551,0.0578784,-0.0381089,0.0660071,-0.03810

89,0.0309332,0.093346,0.0660071,0.093346,-0.0768551,-0.1576958,0.00000

29,0.0000002,0.0000029,0.0848261,0.,0.0000002,0.,0.0848261|Polar=13.81

20382,0.,13.8120382,0.,0.,13.8120382|PG=TD [O(C1),4C3(H1)]|NImag=0||0.

This is a summary of the results printed at the end of each job. One can see "NImag=0" in the bottom right corner. The "NImag" stands for the number of imaginary frequencies. Evidently, there is no imaginary (negative) frequency and therefore, the optimized structure corresponds to a (local) minimum.

THERMODYNAMIC PARAMETERS IN *GAUSSIAN*

The calculation of thermodynamic parameters is very straightforward. These parameters appear as a part of the result of frequency calculations under the "Thermochemistry" section. By default, the calculation is performed at the 298.15 K temperature and 1 atmospheric pressure. However, one can calculate these parameters at any other specified temperature as well.

- Thermochemistry -

Temperature 298.150 Kelvin. Pressure 1.00000 Atm.

Atom 1 has atomic number 6 and mass 12.00000

Atom 2 has atomic number 1 and mass 1.00783

Atom 3 has atomic number 1 and mass 1.00783

Atom 4 has atomic number 1 and mass 1.00783

Atom 5 has atomic number 1 and mass 1.00783

The various thermodynamic parameters are calculated using the partition function (Q). As we know,

$$Q = q_t q_r q_v q_e \tag{14}$$

where q_t, q_r, q_v, and q_e are translational, rotational, vibrational, and electronic contributions in the partition function, respectively. These quantities are obtained as described in any standard text on statistical thermodynamics [10]. By default, the Q is computed considering the bottom (Bot) of the internuclear potential energy well as the zero (reference point) of the energy.

Q Log10(Q) Ln(Q)

Total Bot 0.227635D-12 -12.642760 -29.111032

Total V=0 0.922556D+08 7.964993 18.340074

Vib (Bot) 0.248057D-20 -20.605449 -47.445800

Vib (V=0) 0.100532D+01 0.002304 0.005305

Electronic 0.100000D+01 0.000000 0.000000

Translational 0.252295D+07 6.401908 14.740939

Rotational 0.363731D+02 1.560781 3.593830

Once Q is calculated, you can calculate all thermodynamic parameters of your interest.

Zero-point correction= 0.044803 (Hartree/Particle)

Thermal correction to Energy= 0.047669

Thermal correction to Enthalpy= 0.048613

Thermal correction to Gibbs Free Energy= 0.027486

Sum of electronic and zero-point Energies= -40.483279

Sum of electronic and thermal Energies= -40.480413

Sum of electronic and thermal Enthalpies= -40.479469

Sum of electronic and thermal Free Energies= -40.500596

All these quantities are calculated in atomic units (Hartree). However, some parameters are also obtained in the standard unit (Calories).

E (Thermal) CV S

KCal/Mol Cal/Mol-Kelvin Cal/Mol-Kelvin

Total 29.913 6.444 44.465

Electronic 0.000 0.000 0.000

Translational 0.889 2.981 34.261

Rotational 0.889 2.981 10.122

Vibrational 28.135 0.482 0.082

POPULATION ANALYSIS IN *GAUSSIAN*: MOLECULAR ORBITALS AND ELECTRONIC PARAMETERS

This analysis is performed, by default, automatically with Optimization and/or Frequency or other jobs. This portion of the output is very important, which is used to obtain many electronic parameters/properties.

****

Population analysis using the SCF density.

****

Orbital symmetries:

Occupied (A1) (A1) (T2) (T2) (T2)

Virtual (A1) (T2) (T2) (T2) (T2) (T2) (T2) (A1) (T2) (T2)

(T2) (T2) (T2) (T2) (A1) (A1) (E) (E) (T2) (T2)

(T2) (A1) (T2) (T2) (T2) (T2) (T2) (T2) (A1)

The electronic state is 1-A1.

Alpha occ. eigenvalues -- -10.15133 -0.69745 -0.39634 -0.39634 -0.39634

Alpha virt. eigenvalues -- 0.00991 0.05736 0.05736 0.05736 0.14637

The molecular orbitals are obtained by the LCAO method. One can see that there are five occupied orbitals of CH_4 along with several virtual (unoccupied) orbitals. Their energy values are also given in the atomic units or Hartree (1 Hartree = 27.21 eV). From the above, the energy of the highest occupied molecular orbital (HOMO) is -0.39634 a.u. and that of the lowest unoccupied molecular orbital (LUMO) is 0.00991 a.u.

The energies of HOMO and LUMO, E_{HOMO} and E_{LUMO}, can be used to obtain the ionization potential (*I*) and electron affinity (*A*) using the Koopmans' theorem [11]. Note, however, that *I* and *A* values, thus obtained, are only "approximate" values as the Koopmans' theorem is valid only for HF method. For DFT, the "derivative discontinuity" should be taken into account as suggested by Perdew *et al.* [12]. More accurate values can only be obtained by the difference of energy of molecule with those of its cation and anion, respectively.

$$I = -E_{HOMO} \qquad\qquad (15)$$

$$A = -E_{LUMO} \qquad\qquad (16)$$

Other parameters such as chemical potential (μ), absolute electronegativity (χ), chemical hardness (η) and electrophilic index (ω), *etc.* can be obtained by using the finite-difference methods [13-17]:

$$\chi = \frac{I + A}{2} = -\frac{E_{HOMO} + E_{LUMO}}{2} = -\mu \tag{17}$$

$$\eta = \frac{I - A}{2} = -\frac{E_{HOMO} - E_{LUMO}}{2} = \frac{E_{gap}}{2} \tag{18}$$

where E_{gap} is the frontier orbitals energy gap, which is nothing but the difference of the energies of LUMO and HOMO,

$$E_{gap} = E_{LUMO} - E_{HOMO} \tag{19}$$

$$\omega = \frac{\mu^2}{2\eta} = \frac{\chi^2}{2\eta} \tag{20}$$

These electronic parameters are, sometimes, referred to as molecular descriptors, chemical reactivity descriptors, or global reactivity descriptors. These will be explained and discussed in the succeeding chapters.

POPULATION ANALYSIS IN *GAUSSIAN*: ATOMIC CHARGES AND MULTIPOLE MOMENTS

The next portion of population analysis contains the partial charges on the atoms. By default, these charges are computed using the Mulliken population scheme [18]. Apart from Mulliken atomic charges, atomic polarizability tensor (APT) derived charges [19] may also be listed. The calculation of atomic charges is a unique feature, which can, otherwise, not be measured experimentally.

Mulliken atomic charges:

1

1 C -0.931204

2 H 0.232801

3 H 0.232801

4 H 0.232801

5 H 0.232801

Sum of Mulliken atomic charges = 0.00000

Below atomic charges, there are multiple moments, starting from dipole moment in the units of Debye and higher-order moments. As CH_4 is a symmetric molecule, its dipole moment is zero.

Dipole moment (field-independent basis, Debye):

X= 0.0000 Y= 0.0000 Z= 0.0000 Tot= 0.0000

Quadrupole moment (field-independent basis, Debye-Ang):

XX= -8.5144 YY= -8.5144 ZZ= -8.5144

XY= 0.0000 XZ= 0.0000 YZ= 0.0000

Traceless Quadrupole moment (field-independent basis, Debye-Ang):

XX= 0.0000 YY= 0.0000 ZZ= 0.0000

XY= 0.0000 XZ= 0.0000 YZ= 0.0000

Octapole moment (field-independent basis, Debye-Ang**2):

XXX= 0.0000 YYY= 0.0000 ZZZ= 0.0000 XYY= 0.0000

XXY= 0.0000 XXZ= 0.0000 XZZ= 0.0000 YZZ= 0.0000

YYZ= 0.0000 XYZ= 0.7811

This was all about the information contained in the output (.log) file (Appendix) for the "Opt+Freq" job. This information can also be visualized directly with the help of the *GaussView* using the *Results* menu in the Sketcher panel (see Fig. **4**).

Unlike the .log file which contains readable text, the .chk file can only be read by the *GaussView* program. This file can be used to plot various surfaces and contours, *e.g.*, molecular orbitals, total density, electrostatic potential (ESP), Laplacian, spin density (if applicable), *etc.* Below we display some of these surfaces for CH_4 molecule.

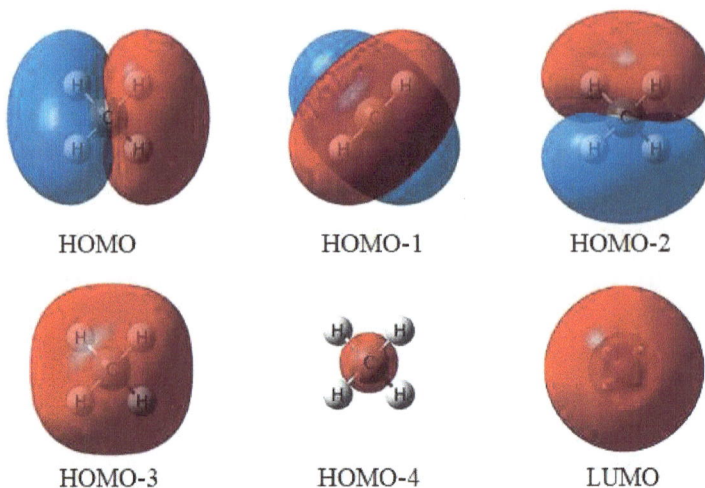

HOMO   HOMO-1   HOMO-2

HOMO-3   HOMO-4   LUMO

Fig. (5). The molecular orbitals' surfaces of CH$_4$ with an isovalue = 0.02.

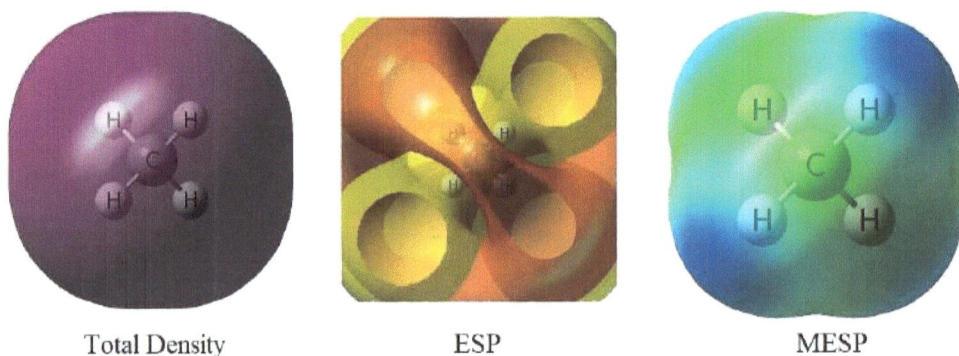

Total Density   ESP   MESP

Fig. (6). Total Density, ESP and MESP surfaces with an isovalue = 0.0004.

The occupied molecular orbitals, starting from the HOMO up to HOMO-4 are displayed in Fig. (**5**) with a default isovalue of 0.02. This isovalue can be adjusted to increase/decrease the surface area for better visualization. It is, however, recommended to use the same isovalue for all molecular orbitals to be plotted. Otherwise, it will be difficult to compare/differentiate the molecular orbitals. For instance, if one increases the isovalue of HOMO-4, it will take the form of HOMO-3 with the default isovalue. Similarly, one can display the LUMO and higher unoccupied molecular orbitals. The LUMO of CH$_4$ is sigma type anti-bonding MO as shown in Fig. (**5**).

Fig. (**6**) shows the total density, ESP, and total density mapped with electrostatic potential (MESP) with a default isovalue of 0.0004. The MESP is most often stands for the molecular electrostatic potential. It is, probably, the *Gaussian* which differentiates ESP and MESP. Almost all other programs just consider MESP. The color-coding in Fig. (**6**) (and also in Fig. **5** to some extent) is very important. The discussion on these surfaces for various molecules can be found in succeeding chapters. Below, we discuss some "explicit" calculations to be performed in *Gaussian*.

NMR CALCULATIONS IN *GAUSSIAN*

The nuclear magnetic resonance (NMR) parameters such as NMR shielding tensors, magnetic susceptibility, spin-spin coupling, *etc.* can be calculated by choosing the "NMR" option as the *Job Type* (see Fig. **4**). The *Gaussian* can perform NMR calculations using various methods such as the Gauge-Independent Atomic Orbital (GIAO) [20, 21], Continuous Set of Gauge Transformations (CSGT) [22], and Individual Gauges for Atoms In Molecules (IGAIM) [23]. The GIAO is the most frequently used, which is employed for NMR calculations discussed in this book.

Actually, the vector potential representing the external magnetic field [24] and consequently, the Hamiltonian and the NMR shielding tensor are not uniquely defined, see eq. (21). The GIAO eliminates the gauge dependence of properties and ensures a uniform description of the molecular system by making the basis functions explicitly dependent on the magnetic field with the inclusion of a complex phase factor that refers to the position of the basis function, which is centered on the nucleus. The CSGT calculations require large basis sets to achieve accurate results. Both CSGT and IGAIM are based on the "atoms in molecule" theory, briefly discussed in Chapter 7 of this book.

The Hamiltonian (H) in an externally applied magnetic field (B_0) for a nuclear spin (I) is expressed as:

$$H = -\frac{\gamma h I \sigma B_0}{2\pi} \tag{21}$$

where γ is gyromagnetic ratio and σ is the NMR shielding tensor (chemical shift tensor), which can be written as:

$$\sigma = \begin{pmatrix} \sigma_{XX} & \sigma_{XY} & \sigma_{XZ} \\ \sigma_{YX} & \sigma_{YY} & \sigma_{YZ} \\ \sigma_{ZX} & \sigma_{ZY} & \sigma_{ZZ} \end{pmatrix} \tag{22}$$

The isotropic and anisotropic components of σ can be obtained as below:

$$\sigma_{iso} = \frac{\sigma_{XX} + \sigma_{YY} + \sigma_{ZZ}}{3} \tag{23}$$

$$\sigma_{anis} = \frac{(\sigma_{YY} - \sigma_{XX})}{2} \tag{24}$$

Note that the molecular structures used for NMR calculations should have been optimized. Below is the relevant part of NMR output file:

Calculating GIAO nuclear magnetic shielding tensors.

SCF GIAO Magnetic shielding tensor (ppm):

1 N Isotropic = 66.6765 Anisotropy = 87.2620

XX= 34.0412 YX= -35.4273 ZX= 73.3530

XY= -17.3319 YY= 95.3218 ZY= 18.3511

XZ= 66.4511 YZ= 28.0007 ZZ= 70.6666

Eigenvalues: -29.7299 104.9083 124.8512

2 H Isotropic = 22.4386 Anisotropy = 5.5879

XX= 22.4714 YX= -0.2869 ZX= -3.0732

XY= 0.3396 YY= 24.9408 ZY= -2.3952

XZ= -3.3369 YZ= -1.6840 ZZ= 19.9038

Eigenvalues: 17.3560 23.7960 26.1639

3 C Isotropic = 66.4423 Anisotropy = 163.4395

XX= 43.0913 YX= -63.6504 ZX= -66.7630

XY= -66.4714 YY= 86.8006 ZY= 34.4632

XZ= -65.4220 YZ= 29.0639 ZZ= 69.4349

Eigenvalues: -22.5376 46.4624 175.4019

One can note the components of the σ matrix and its eigenvalues as per eq. (22), (23) and (24).

However, the chemical shift (δ) is usually, expressed in parts per million (ppm) as below:

$$\delta = 10^6 \, (\sigma_{\text{ref}} - \sigma_{\text{mol}}) \tag{25}$$

Here σ_{ref} is the NMR shielding values for reference molecule, which is generally tetramethylsilane (TMS), $Si(CH_3)_4$. TMS is inert, volatile, and soluble in most organic solvents. More importantly, it has 12 equivalent ^1H and 4 equivalent ^{13}C nuclei. Note that the NMR spectrum is recorded for ^1H and ^{13}C nuclei, experimentally. The discussion on the NMR chemical shifts for several molecules can be found in the next few chapters.

NLO PARAMETERS IN *GAUSSIAN*

The nonlinear optical (NLO) parameters such as polarizability and hyperpolarizability, *etc.* can be calculated by including "Polar" as an *Additional Keywords* in *Gaussian* (see Fig. **4**). The total energy of a molecule changes, when it is placed under a static electric field, and this change can be calculated as [25],

$$E = E^0 - \mu_i F_i - \frac{1}{2}\alpha_{ij}F_iF_j - \frac{1}{6}\beta_{ijk}F_iF_jF_k - \dots \tag{26}$$

where E^0 is the total energy in the absence of an electric field; F_i represents the components of the electric field, α_{ij} is a second rank tensor called the polarizability tensor and β_{ijk} is the first in an infinite series of dipole hyperpolarizabilities.

The mean values of dipole moment (μ_0), polarizability (α_0), and first hyperpolarizability (β_0) can be obtained by numerical differentiation with a static electric field magnitude of 0.001 a.u. using the finite-field approach [26] as below:

$$\mu_0 = (\mu_x^{\,2} + \mu_y^{\,2} + \mu_z^{\,2})^{\frac{1}{2}} \tag{27}$$

$$\alpha_0 = \frac{1}{3}(\alpha_{xx} + \alpha_{yy} + \alpha_{zz}) \tag{28}$$

$$\beta_0 = \left[\beta_x^{\,2} + \beta_y^{\,2} + \beta_z^{\,2} \right]^{\frac{1}{2}} \tag{29}$$

The β_x, β_y, and β_z can be calculated using the tensor components as follows,

$$\beta_i = \frac{3}{5}\left(\beta_{iii} + \beta_{ijj} + \beta_{ikk}\right) \tag{30}$$

The components of dipole moment and total (mean) dipole moment can be found as below:

Electric dipole moment (input orientation):

(Debye = 10**-18 statcoulombcm, SI units = C m)

(au) (Debye) (10**-30 SI)

Tot 0.232385D+01 0.590665D+01 0.197024D+02

x -0.357857D-01 -0.909583D-01 -0.303404D+00

y -0.787666D-01 -0.200205D+00 -0.667811D+00

z 0.232224D+01 0.590255D+01 0.196888D+02

Similarly, the components of polarizability and its isotropic (mean) value can be found here:

Dipole polarizability, Alpha (input orientation).

(esu units = cm**3, SI units = C**2 m**2 J**-1)

Alpha(0;0):

(au) (10**-24 esu) (10**-40 SI)

iso 0.604279D+04 0.895449D+03 0.996321D+03

aniso 0.783998D+04 0.116177D+04 0.129264D+04

xx 0.865667D+04 0.128279D+04 0.142729D+04

yx 0.537350D+02 0.796271D+01 0.885971D+01

yy 0.865481D+04 0.128251D+04 0.142698D+04

zx -0.217398D+02 -0.322151D+01 -0.358441D+01

zy -0.506208D+02 -0.750122D+01 -0.834624D+01

zz 0.816890D+03 0.121051D+03 0.134687D+03

Likewise, the first (dipole) hyperpolarizability and their various components can be obtained as follows:

First dipole hyperpolarizability, Beta (input orientation).

||, _|_ parallel and perpendicular components, (z) with respect to z axis,

vector components x,y,z. Values do not include the 1/n! factor of 1/2.

(esu units = statvolt**-1 cm**4, SI units = C**3 m**3 J**-2)

Beta(0;0,0):

(au) (10**-30 esu) (10**-50 SI)

|| (z) -0.715809D+04 -0.618403D+02 -0.229514D+02

_|_(z) -0.238603D+04 -0.206134D+02 -0.765047D+01

x -0.597157D+05 -0.515897D+03 -0.191470D+03

y -0.656141D+06 -0.566855D+04 -0.210383D+04

z -0.357904D+05 -0.309202D+03 -0.114757D+03

|| 0.131965D+06 0.114007D+04 0.423127D+03

xxx -0.721638D+05 -0.623439D+03 -0.231383D+03

xxy -0.290897D+05 -0.251312D+03 -0.932721D+02

yxy 0.291466D+05 0.251804D+03 0.934547D+02

yyy -0.158353D+06 -0.136805D+04 -0.507738D+03

xxz -0.112853D+06 -0.974958D+03 -0.361846D+03

yxz -0.383942D+05 -0.331696D+03 -0.123106D+03

yyz -0.163028D+06 -0.140843D+04 -0.522726D+03

zxz 0.231119D+05 0.199669D+03 0.741052D+02

zyz -0.312705D+05 -0.270153D+03 -0.100265D+03

zzz 0.263950D+06 0.228032D+04 0.846320D+03

More conveniently, these values are also summarized at the end of the output file. The extract of the summary reads:

Dipole=-0.0357857,-0.0787666,2.3222422\Polar=8656.6732

672,53.7350428,8654.8062436,-21.7398028,-50.6207697,816.8903216\HyperP

olar=-72163.8017394,-29089.6911174,29146.6400809,-158353.4762657,-1128

52.5012179,-38394.1766942,-163027.8496637,23111.9355625,-31270.5056575

,263950.2024716\

Here the components of *Dipole* moment, *Polar*izability, and (first) *HyperPolar*izability appear in the same order as described above. Note that these values are for static cases (frequency = 0), which will be discussed in the next few chapters. However, frequency-dependent NLO parameters may be calculated by including "CPHF=RdFreq" along with the "Polar" and then specifying the frequency as an *Add*itional *Inp*ut (see Fig. **4**).

NBO ANALYSIS IN *GAUSSIAN*

The natural bond orbital (NBO) calculations in *Gaussian* can be initiated by the "pop=nbo" as *Additional Keywords*. This calculation is actually performed by an external program NBO 3.1 [27] as implemented in the *Gaussian 09*. The NBO program tries to find the Lewis structure closest to the given structure (Of course, this needs an optimized structure). One can use "pop=nbo" in combination with "Opt+Freq" job, however, the results of NBO calculations performed after optimization should be considered. The NBO calculations start from here:

***************Gaussian NBO Version 3.1***********************

NATURAL ATOMIC ORBITAL AND NATURAL BOND ORBITAL ANALYSIS

***************Gaussian NBO Version 3.1***********************

/RESON /: Allow strongly delocalized NBO set

Analyzing the SCF density

Job title: Title Card Required

Storage needed: 3724 in NPA, 4773 in NBO (33554324 available)

The natural population analysis (NPA) [28] is the first step in NBO calculations. The NPA is similar to Mulliken population analysis, which is used to obtain natural atomic orbitals (NAOs) [29].

NATURAL POPULATIONS: Natural atomic orbital occupancies

NAO Atom No langType(AO) Occupancy Energy

----

1 C 1 S Cor(1S) 1.99953 -10.01040

2 C 1 S Val(2S) 1.15931 -0.28654

3 C 1 S Ryd(4S) 0.00000 0.78300

4 C 1 S Ryd(3S) 0.00000 0.18420

5 C 1 S Ryd(5S) 0.00000 23.32067

6 C 1 px Val(2p) 1.21147 -0.09493

7 C 1 px Ryd(3p) 0.00000 0.66178

8 C 1 px Ryd(4p) 0.00000 0.72780

9 C 1 px Ryd(5p) 0.00000 2.08949

The all possible orbitals (s, p, d, *etc.*) of each type core (Cor), valence (Val), Rydberg (Ryd) for each atom are listed along with their energies and occupancies. Next appears the summary of natural population analysis:

Summary of Natural Population Analysis:

Natural Population

Natural --

Atom No Charge Core Valence Rydberg Total

C 1 -0.79663 1.99953 4.79372 0.00339 6.79663

H 2 0.19916 0.00000 0.80061 0.00023 0.80084

H 3 0.19916 0.00000 0.80061 0.00023 0.80084

H 4 0.19916 0.00000 0.80061 0.00023 0.80084

H 5 0.19916 0.00000 0.80061 0.00023 0.80084

===
=======

* Total * 0.00000 1.99953 7.99617 0.00431 10.00000

One can find the Natural Charge (often referred to as NPA or NBO charge), which is considered to be more accurate than the Mulliken charge due to the reason that the NPA assigns the maximum possible occupancy to each NAO. After the NPA, the NBO analysis [30] starts:

NATURAL BOND ORBITAL ANALYSIS:

Occupancies Lewis Structure Low High

Occ. ------------------- ----------------- occ occ

Cycle Thresh. Lewis Non-Lewis CR BD3C LP (L) (NL) Dev

===
=========

1(1)1.90 9.99798 0.00202 1 4 0 0 0 0 0.00

The NBOs are then formed using the NAOs and the following information is obtained:

(Occupancy) Bond orbital/ Coefficients/ Hybrids

----

1. (1.99961) BD (1) C 1 - H 2

(59.97%) 0.7744* C 1 s(25.00%)p 3.00(74.93%)d 0.00(0.07%)

0.0000 0.5000 0.0000 0.0000 0.0000

0.4998 0.0000 0.0000 0.0000 0.4998

0.0000 0.0000 0.0000 0.4998 0.0000

0.0000 0.0000 0.0153 0.0153 0.0153

0.0000 0.0000

(40.03%) 0.6327* H 2 s(100.00%)

1.0000 -0.0003 0.0002

This can be interpreted as a single bonding orbital for C1-H2 with 1.99961 electrons has 59.97% C1 character in an s0.25 p3 hybrid and has 40.41% H_2 character in an s orbital. Similarly, the composition of other NBOs is listed such as core (CR), Rydberg (RY), lone-pair (LP), *etc.* The corresponding anti-bonds are denoted as an asterisk (*).

Below this, the most important portion of the NBO analysis. It is the portion, which will be discussed in a few succeeding chapters. The localized (Lewis) NBOs can interact strongly. A filled bonding or lone pair orbital can act as a donor and an empty antibonding or lone pair orbital can act as an acceptor. These interactions can strengthen and weaken bonds. The calculation takes all possible interactions between `filled' (donor) Lewis-type NBOs and `empty' (acceptor) non-Lewis NBOs, and estimating their energetic importance by second-order perturbation theory. The results are printed in the following format:

Second Order Perturbation Theory Analysis of Fock Matrix in NBO Basis

Threshold for printing: 0.50 kcal/mol

E(2) E(j)-E(i) F(i,j)

Donor NBO (i) Acceptor NBO (j) kcal/mol a.u. a.u.

===
=========

Since these interactions lead to loss of occupancy from the localized NBOs of the ideal Lewis structure into the non-Lewis orbitals, they are referred to as `delocalization' corrections to the natural Lewis structure. The energy ($E^{(2)}$) corresponding to the interaction between the ith donor and j^{th} acceptor NBOs is obtained as below:

$$E^{(2)} = -n_i \frac{F_{ij}^{\,2}}{E_j - E_i} \tag{31}$$

where F_{ij} is the Fock matrix element between the ith and j^{th} NBOs having energies E_i and E_j, respectively, and n_i is the population of ith NBO. More about NBO can be found here [31].

TDDFT CALCULATION OF UV-VIS-NIR SPECTRUM IN *GAUSSIAN*

So far, all the calculations performed were in the ground state of the molecule. The electronic transitions resulting in absorption and emission spectra fall in the UV-Vis-NIR region. This spectrum is obtained by excited-state calculations using the TDDFT method [32], as discussed in the previous chapter. This calculation is performed using "TD-SCF" instead of the *Ground State* option in the *Method* tab (see Fig. **4**). By default, *Gaussian* calculates only up to 3 excited states, however, more excited states can be included by explicit mention. The output of TDDFT reads as below:

****

Excited states from <AA,BB:AA,BB> singles matrix:

****

Ground to excited state transition densities written to RWF 633

Ground to excited state transition electric dipole moments (Au):

state X Y Z Dip. S. Osc.

1 -0.1236 0.0268 -0.0341 0.0172 0.0019

2 0.3463 0.2434 -0.4499 0.3816 0.0460

3 0.6788 -0.4377 -0.0356 0.6536 0.0802

4 0.8258 -0.8479 -0.2441 1.4605 0.1810

5 -1.3539 0.3873 0.1904 2.0193 0.2543

6 0.0469 -0.1501 -0.3841 0.1723 0.0221

The TDDFT first calculates the transition electric dipole moments, their components (X, Y, and Z), and oscillator strength for all the excited states (six in this case). The calculation gives, then, the components of transition velocity, transition magnetic dipole moments, *etc.* in the same format.

The oscillator strength (*f*) is an important factor for the measurement of absorptivity as well as the intensity of an electronic transition, *i.e.*, how strongly the particular electronic transition is allowed. It is a dimensionless quantity defined by the formula,

$$f = \frac{4 m_e c \varepsilon_0 B}{N_A e^2} \tag{32}$$

where m_e is the mass of the electron, ε_0 is the vacuum permittivity, N_A is Avogadro constant, e is the elementary charge and B is the molar natural absorption coefficient integrated over the whole band units of the frequency. The oscillator strength is a number usually lying between zero and one. Forbidden transitions have oscillator strengths close to zero.

The most important part of the output appears as below:

Excitation energies and oscillator strengths:

Excited State 1: Singlet-A 4.6222 eV 268.23 nm f=0.0019 <S**2>=0.000

56 -> 59 0.13581

56 -> 60 0.52657

56 -> 61 -0.34440

57 -> 60 0.23620

57 -> 61 -0.14582

This state for optimization and/or second-order correction.

Total Energy, E(TD-HF/TD-KS) = -1041.25890730

Copying the excited state density for this state as the 1-particle RhoCI density.

Excited State 2: Singlet-A 4.9203 eV 251.98 nm f=0.0460 <S**2>=0.000

55 -> 60 0.14830

55 -> 61 0.17461

56 -> 59 -0.13489

57 -> 59 0.52524

58 -> 59 0.35797

The excitation (transition) energy (in eV), excitation wavelength (in nm), and oscillator strengths (f) are printed for each transition state. The peak(s) in the UV-Vis-NIR spectrum corresponds to the maximum f value(s) and the corresponding wavelength is designated as the maximum excitation wavelength (λ_{max}). The electronic-transition coefficients (a) between molecular orbitals (whose numbers are given) corresponding to every excited are also listed. These coefficients can be used to determine the transition probability (T) as below:

$$T = 2a^2 \tag{33}$$

For instance, in the 1st excited-state, the probability of MO(56)→MO(60) transition is 0.5545 or (55.45%).

Note that the TDDFT calculations require the "Opt" keyword for the accurate results, which is very expensive in terms of computational cost. Therefore, a reasonably optimized geometry in the ground may be used for single-point "Energy" calculations using the TDDFT method. More information about TDDFT calculations can be found here [33].

CONCLUDING REMARKS

In this chapter, we discussed the basics of geometry optimization. It was pointed out that the "Frequency" calculations MUST be performed after the "Optimization" in order to obtain "truly" optimum structures of molecular systems. We have provided a brief introduction of the *Gaussian* and *GaussView* programs, which are widely used for DFT calculations worldwide. We have provided the interpretation of *Gaussian* output for the "Opt+Freq" job. The various surfaces such as molecular orbitals, electron density and MESP have been also plotted for completeness. In addition, we have explicitly discussed the NMR, NLO, NBO, and TDDFT calculations in the *Gaussian*. We believe that this chapter will be helpful for beginners to perform the DFT calculations and understand the results obtained. Further, this chapter will help the readers to grasp the discussion in the succeeding chapters.

CONSENT FOR PUBLICATION

Not applicable.

CONFLICT OF INTEREST

The authors declare no conflict of interest, financial or otherwise.

ACKNOWLEDGEMENTS

The author acknowledges the funding from University Grant Commission, India through Startup grant [30-466/2019(BSR)].

REFERENCES

[1]   Bernhard, S. *Infrared & Raman Spectroscopy*; Wiley VCH Inc: New York, **1995**.

[2]   Arnett, E.M.; Larsen, J.W. Stabilities of carbonium ions in solution. VI. Large Baker-Nathan effect for alkylbenzenonium ions. *J. Am. Chem. Soc.*, **1969**, *91*, 1438-1442.
[http://dx.doi.org/10.1021/ja01034a029]

[3]   https://www.msg.chem.iastate.edu/

[4]   http://classic.chem.msu.su/gran/gamess/index.html

[5]   https://www.faccts.de/orca/

[6]   Frisch, M.J.; Trucks, G.W.; Schlegel, H.B. *Gaussian 09, Revision B.01*; Gaussian Inc: Wallingford, CT, **2010**.

[7]   http://gaussian.com/

[8]   https://gaussian.com/keywords/

[9]   http://gaussian.com/iop/

[10]   McQuarrie, D.A.; Simon, J.D. *Molecular Thermodynamics*; University Science Books: USA, **1999**.

[11]   Koopmans, T. Uber die Zuordnung von Wellenfunktionen und Eigenwerten zu den einzelnen Elektronen eines Atoms. *Physica,* **1934**, *1*, 104-113.
[http://dx.doi.org/10.1016/S0031-8914(34)90011-2]

[12]   Perdew, J.P.; Parr, R.G.; Levy, M.; Balduz, J.L., Jr Density-Functional Theory for Fractional Particle Number: Derivative Discontinuities of the Energy. *Phys. Rev. Lett.*, **1982**, *49*, 1691-1694.
[http://dx.doi.org/10.1103/PhysRevLett.49.1691]

[13]   Pearson, R.G. Absolute electronegativity and absolute hardness of Lewis acids and bases. *J. Am. Chem. Soc.*, **1985**, *107*, 6801-6806.
[http://dx.doi.org/10.1021/ja00310a009]

[14]   Pearson, R.G.; Dillon, R.L. Rates of ionization of pseudo acids. IV. relation between rates and equilibria. *J. Am. Chem. Soc.*, **1953**, *75*, 2439-2443.
[http://dx.doi.org/10.1021/ja01106a048]

[15]   Parr, R.G.; Pearson, R.G. Absolute hardness: companion parameter to absolute electronegativity. *J. Am. Chem. Soc.*, **1983**, *105*, 7512-7516.
[http://dx.doi.org/10.1021/ja00364a005]

[16]   Parr, R.G.; Szentpaly, L.V.; Liu, S. Electrophilicity index. *J. Am. Chem. Soc.*, **1999**, *121*, 1922-1924.
[http://dx.doi.org/10.1021/ja983494x]

[17]   Chattaraj, P.K.; Lee, H.; Parr, R.G. HSAB principle. *J. Am. Chem. Soc.*, **1991**, *113*, 1855-1856.
[http://dx.doi.org/10.1021/ja00005a073] [PMID: 12603158]

[18]   Mulliken, R.S. Electronic population analysis on LCAO–MO molecular wave functions. I. *J. Chem. Phys.*, **1955**, *23*, 1833-1840.
[http://dx.doi.org/10.1063/1.1740588]

[19]   Cioslowski, J. A new population analysis based on atomic polar tensors. *J. Am. Chem. Soc.*, **1989**, *111*, 8333-8336.
[http://dx.doi.org/10.1021/ja00204a001]

[20]   Ditchfield, R. Self-consistent perturbation theory of diamagnetism. 1. Gauge-invariant LCAO method for N.M.R. chemical shifts. *Mol. Phys.,* **1974**, *27*, 789-807.
[http://dx.doi.org/10.1080/00268977400100711]

[21]   Wolinski, K.; Hilton, J.F.; Pulay, P. Efficient Implementation of the Gauge-Independent Atomic Orbital Method for NMR Chemical Shift Calculations. *J. Am. Chem. Soc.,* **1990**, *112*, 8251-8260.
[http://dx.doi.org/10.1021/ja00179a005]

[22]   Keith, T.A.; Bader, R.F.W. Calculation of magnetic response properties using a continuous set of gauge transformations. *Chem. Phys. Lett.,* **1993**, *210*, 223-231.
[http://dx.doi.org/10.1016/0009-2614(93)89127-4]

[23]   Keith, T.A.; Bader, R.F.W. Calculation of magnetic response properties using atoms in molecules. *Chem. Phys. Lett.,* **1992**, *194*, 1-8.
[http://dx.doi.org/10.1016/0009-2614(92)85733-Q]

[24]   Griffiths, D.J. *Introduction to Electrodynamics*; Pearson: USA, **2013**.

[25]   Buckingham, A.D. Permanent and induced molecular moments and long-range intermolecular forces. *Adv. Chem. Phys.,* **1967**, *12*, 107-142.
[http://dx.doi.org/10.1002/9780470143582.ch2]

[26]   Cohen, H.D.; Roothaan, C.C.J. Electric-dipole polarizability of atoms by the hartree-fock method. I. Theory for closed shell systems. *J. Chem. Phys.,* **1965**, *43*, 534-539.
[http://dx.doi.org/10.1063/1.1696518]

[27]   Glendening, E.D.; Badenhoop, J.K.; Reed, A.E.; Carpenter, J.E.; Weinhold, F. *NBO 3.1 Program*; Theoretical Chemistry Institute, University of Wisconsin: Madison, WI, **1996**.

[28]   Reed, A.E.; Weinstock, R.B.; Weinhold, F. Natural population analysis. *J. Chem. Phys.,* **1985**, *83*, 735-746.
[http://dx.doi.org/10.1063/1.449486]

[29]   Reed, A.E.; Weinhold, F. Natural bond orbital analysis of near Hartree-Fock water dimer. *J. Chem. Phys.,* **1983**, *78*, 4066-4073.
[http://dx.doi.org/10.1063/1.445134]

[30]   Foster, J.P.; Weinhold, F. Natural hybrid orbitals. *J. Am. Chem. Soc.,* **1980**, *102*, 7211-7218.
[http://dx.doi.org/10.1021/ja00544a007]

[31]   Reed, A.E.; Curtiss, L.A.; Weinhold, F. Intermolecular interactions from a natural bond orbital, donor-acceptor viewpoint. *Chem. Rev.,* **1988**, *88*, 899-926.
[http://dx.doi.org/10.1021/cr00088a005]

[32]   Runge, E.; Gross, E.K.U. Density-functional theory for time-dependent systems. *Phys. Rev. Lett.,* **1984**, *52*, 997-1000.
[http://dx.doi.org/10.1103/PhysRevLett.52.997]

[33]   Adamo, C.; Jacquemin, D. The calculations of excited-state properties with Time-Dependent Density Functional Theory. *Chem. Soc. Rev.,* **2013**, *42*(3), 845-856.
[http://dx.doi.org/10.1039/C2CS35394F] [PMID: 23117144]

CHAPTER 3

DFT Study on Some Synthetic Compounds: (2,6), (2,4) and (3,4) Dichloro Substituted Phenyl-N-(1-3-thiazol-2-yl) Acetamides

Abstract: In this chapter, we present and discuss DFT study on three dichloro substituted (1, 3-thiazol-2-yl) acetamides; 26DTA, 24DTA and 34DTA using the B3LYP/6-31+G(d,p) method. We focus on the need of scaling the normal modes of vibrations and test two scaling schemes on 26DTA. We analyze their performance by comparing the scaled values against FTIR data. Subsequently, a detailed comparative study on the spectroscopic properties of 24DTA and 34DTA has been performed using a better scaling scheme. In addition, the NBO analysis is employed to explain the coordination ability of molecules and several electronic parameters are obtained to describe their chemical reactivity. This chapter is expected to provide the first flavor of the real application of DFT on biologically active molecules.

Keywords: Acetamide, B3LYP, Coordination ability, DFT, Electron parameters, FTIR, HOMO, LUMO, NBO, NPA, Scaling equation, Scaling factor, Vibrational spectra.

INTRODUCTION

Acetamide based heterocycles having nitrogen and sulphur atoms, form a distinct group of pharmacologically important compounds with an amide bond similar to that between amino acids in proteins (Chapter 4 will specifically address unusual amino acids). Several natural products and drug compounds contain such heterocyclic moieties [1, 2], including the derivatives of thiazole, which have several biological activities [3, 4]. For instance, the thiazoles have been used in the development of drugs for the cure of several diseases [5 - 12], fibrinogen receptor antagonists with antithrombotic activity [13] an inhibitor of bacterial DNA gyrase B [14]. It is, therefore, desirable to explore some novel compounds based on the thiazole ring. In this chapter, we present DFT based studies on three molecules, namely, 2-(2,6-Dichlorophenyl)-N-(1,3-thiazol-2-yl) acetamide [26DTA], 2-(2,4-Dichlorophenyl)-N-(1,3-thiazol-2-yl) acetamide [24DTA], and 2- (3,4-Dichlorophenyl)-N-(1,3-thiazol-2-yl) acetamide [34DTA]. The results [15 - 17] were obtained by the B3LYP functional and 6-31+G(d,p) basis set.

Ambrish Kumar Srivastava and Neeraj Misra

MOLECULAR GEOMETRIES

The crystal structures obtained by X-Ray diffraction (XRD) of these compounds can be found in the literature [18 - 20], which were employed to generate the input structure of these molecules to be used for geometry optimization. The molecular structures, after optimization, of 26DTA, 24DTA, and 34DTA are displayed in Fig. (1) and some structural parameters can be found in Table 1. The C–C bond lengths in the vicinity of Cl atoms become a bit longer than others. The bond angle C6-C1-C2 is decreased by 3-4⁰ in molecules 26DTA and 24DTA having at least one substitution of Cl at *ortho* position. The positional change in the substitutions leads to the decrease in the dihedral C1-C2-C7-C8 continuously from molecule 24DTA to 34DTA. In the crystal phase [18 - 20], the existence of N···H–N hydrogen bonds modifies their structural parameters. These inter-molecular interactions can't be considered by performing calculations on an isolated molecule in the gas phase. These molecules also possess intra-molecular hydrogen bonds (see Chapter 7).

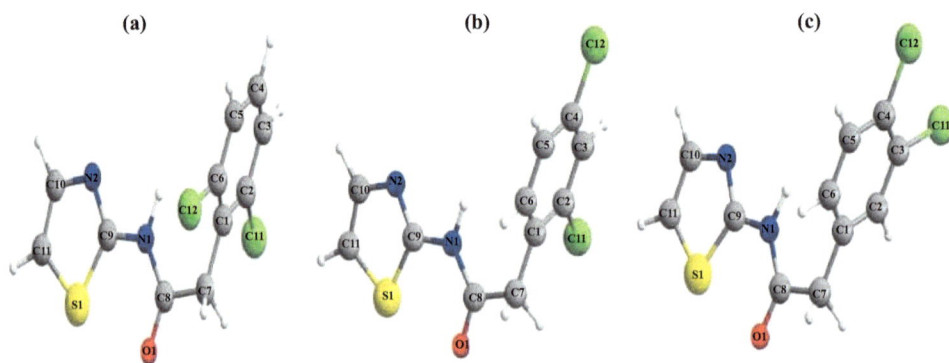

Fig. (1). Molecular structures of **(a)** 26DTA, **(b)** 24DTA, and **(c)** 34DTA at the B3LYP/6-31+G(d,p) level [17].

Table 1. Selected structural parameters at the B3LYP/6-31+G(d,p) level [17]. The corresponding experimental data are taken from Ref. [18-20].

Structural Parameters (Å, °)	26DTA		24DTA		34DTA	
	Calc.	Expt. [18]	Calc.	Expt. [19]	Calc.	Expt. [20]
C1-C2	1.409	1.400	1.406	1.391	1.400	1.398
C2-C3	1.394	1.390	1.395	1.393	1.397	1.389
C3-C4	1.393	1.386	1.394	1.386	1.401	1.392
C4-C5	1.393	1.385	1.394	1.384	1.397	1.390
C1-C6	1.409	1.402	1.404	1.400	1.403	1.391
C1-C7	1.509	1.512	1.509	1.504	1.511	1.511

(Table 1) cont.....

C8-O1	1.224	1.226	1.224	1.227	1.225	1.225
C8-N1	1.371	1.363	1.372	1.368	1.371	1.365
C9-N2	1.306	1.311	1.305	1.315	1.305	1.315
C9-S1	1.752	1.730	1.752	1.727	1.752	1.720
C-Cl1	1.759	1.743	1.756	1.744	1.745	1.736
C-Cl2	1.759	1.738	1.752	1.741	1.744	1.728
C6-C1-C2	115.7	115.6	116.8	117.2	118.3	118.7
C5-C4-C3	120.2	120.3	121	122.2	119.3	119.3
C6-C1-C7	122.1	122.4	120.5	120.8	121	122.3
C1-C7-C8	118	110.3	117.7	113	117.9	111.9
C8-N1-C9	125.7	123.2	125.7	123.9	125.7	122.9
N1-C9-N2	120.4	121.1	120.4	120.6	120.4	120.8
C9-N2-C10	110.1	109.5	110.1	109.7	110.1	109.2
C9-S1-C11	87.7	88.8	87.7	88.7	87.7	88.7
C6-C1-C7-C8	90.4	91.89	100.3	112.4	86.8	110.9
C1-C7-C8-N1	0	125.3	14.8	172.5	3.7	-166.9

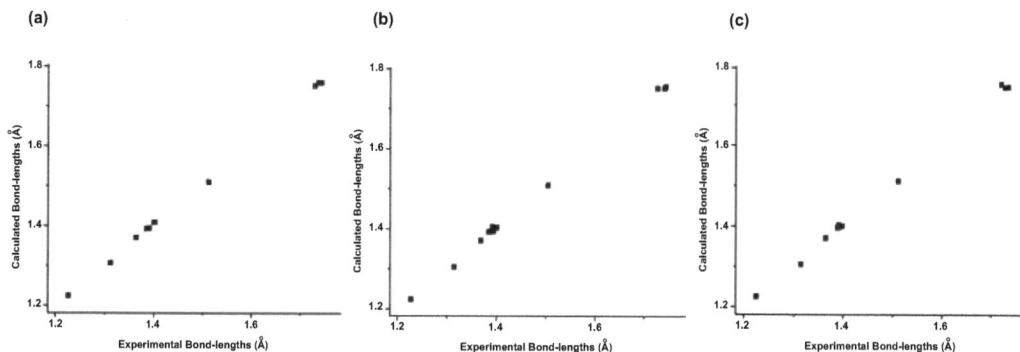

(a)   **(b)**   **(c)**

Fig. (2). Correlation between calculated and experimental bond lengths; **(a)** $d_{calc.} = 1.040\, d_{expt.} - 0.051$, $R^2 = 0.9991$ for 26DTA, **(b)** $d_{calc.} = 1.035\, d_{expt.} - 0.044$, $R^2 = 0.9986$ for 24DTA, and **(c)** $d_{calc.} = 1.039\, d_{expt.} - 0.050$, $R^2 = 0.9980$ for 34DTA.

In Table **1**, we have also listed the bond length obtained from XRD on these compounds for comparison. In Fig. (**2**), we have plotted correlation graphs between calculated and experimental bond-lengths, excluding the C-H bond-lengths. We obtain a nice linear correlation having a correlation coefficient (R^2) of 0.9980-0.9991, which verifies that the present computational method is quite appropriate to reproduce the experimental geometry.

MOLECULAR ORBITALS AND ELECTRONIC PROPERTIES

The highest occupied molecular orbital (HOMO) determines the ability to donate, whereas the lowest unoccupied molecular orbital (LUMO) represents the ability to accept an electron. The HOMO and LUMO surfaces for these molecules are plotted in Fig. (3). For all molecules, the HOMO is localized on the thiazole ring as well as the amide fragment and LUMO is contributed primarily by the phenyl ring. Evidently, the HOMO→LUMO transition in these molecules suggests the transfer of charge to the phenyl ring. The smaller E_{gap} (see Chapter 2), 4.89 eV of 26DTA and 24DTA as compared to 4.94 eV in 34DTA indicates that they are chemically more reactive than the 34DTA. One may, therefore, conclude that two Cl atoms attached to neighboring carbon atoms of the phenyl ring in 34DTA make it less reactive.

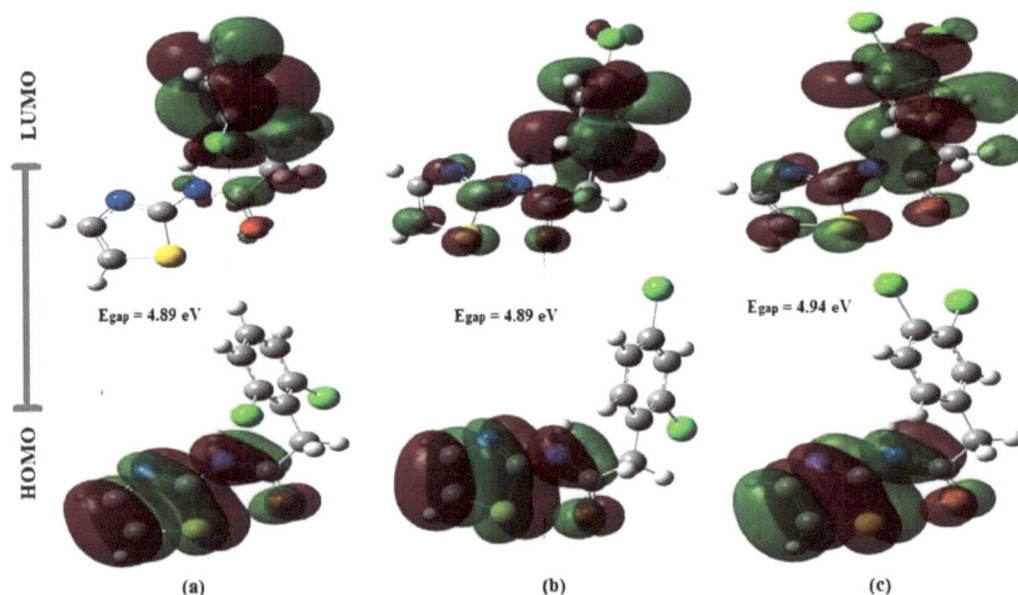

Fig. (3). HOMO and LUMO plots of **(a)** 26DTA, **(b)** 24DTA, and (c) 34DTA.

The HOMO and LUMO energies can be employed to obtain some electronic parameters describing the chemical reactivity of the molecule as mentioned in the previous chapter (see Chapter 2). These parameters, namely, ionization potential (I), electron affinity (A), absolute electro-negativity (χ), and chemical hardness (η) along with the dipole moment (μ) of 26DTA, 24DTA, and 34DTA can be found in Table **2**.

One can see that the electronic parameters change with the change in the position of substitutions. 26DTA has smaller I and A values, which become slightly larger for 34DTA. Therefore, the chemical reactivity depends on the positions of substitution. The greater electronegativity of 34DTA might indicate its higher coordination ability, but larger chemical hardness suggests its relatively higher kinetic stability. Interestingly, one can see that the dipole moments adopt an opposite trend of I and A values. Larger dipole moment stands with the smaller I and A values. The higher dipole moment of the 26DTA molecule reveals that it is relatively more polarized, whereas 34DTA is less polarized.

Table 2. Electronic parameters of 26DTA, 24DTA, and 34DTA [17].

Molecule	I (eV)	A (eV)	χ (eV)	η (eV)	μ (Debye)
26DTA	6.37	1.48	3.92	2.44	3.784
24DTA	6.43	1.54	3.98	2.44	2.165
34DTA	6.49	1.55	4.02	2.47	1.272

NBO ANALYSIS

Charge Distribution

The NPA charges of 26DTA, 24DTA, and 34DTA are graphically presented in Fig. (**4**). In 24DTA, these charges range from -0.655|e| to 0.690|e|. In 26DTA and 34DTA, however, this value ranges from -0.653|e| to 0.698|e| and from -0.649|e| to 0.690|e|, respectively. In all three molecules, the maximum charge on C8 atom was seen due to the effect of positively charged carbon while maximum negative charge is contained by N1 of amide fragment. The negative charge on N1 in 34DTA is less than that in 26DTA or 24DTA. This implies that the charge transfer from N1 to thiazole ring is enhanced in 34DTA. The carbon atoms of the phenyl ring, are all negatively charged. Negative charges are, however, reduced at the carbon sites connected with Cl atoms. Thus, in the presence of Cl atoms, all C atoms accept electrons.

Coordination Ability

Amides have been widely known for their coordinating ability, which facilitates them to be employed as ligands [21]. To study the coordination ability of these compounds, we have carried out the NBO analyses. The results obtained by the NBO calculations have been employed to analyze the second-order interaction between suitable lone-pair (LP) orbitals (n) and appropriate antibonding orbitals (σ^*) and determine the lone-pair localization and consequently, the possible coordination ability. The one-centre valence LPs have been found to be the most suitable in coordinating with the metal ions.

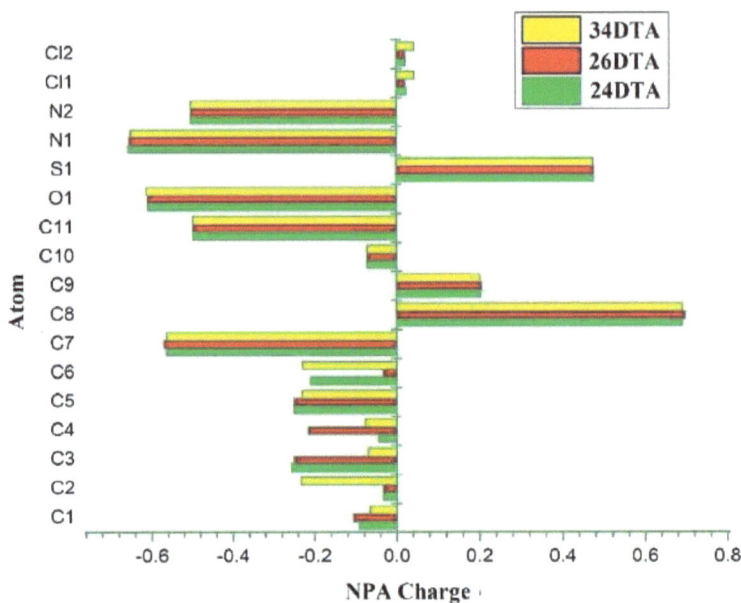

Fig. (4). NPA charges of 26DTA, 24DTA, and 34DTA [17].

The parameters obtained by the NBO analysis are listed in Table **3**. We studied the interactions of LPs of oxygen (O1) and nitrogen (N1) with various antibonding orbitals. These LPs possess primarily p character, which are generally occupied by a pair of electrons. In the interactions of LP and antibonds, LP donates its occupancy to antibond. The stabilisation energy, $E^{(2)}$ corresponding to these interactions is employed to estimate the engagement of the lone pairs in the intra-molecular delocalization. The higher $E^{(2)}$ values show the stronger interaction between LPs (donor) and antibonding orbitals (acceptor). The interactions of LP of N1 and antibonds, *i.e.*, $n_1(N1) \to \pi^*(C8\text{-}O1)$ and $n_1(N1) \to \pi^*(C9\text{-}N2)$ result in the significant stabilization by an amount of 61 and 42 kcal/mol in 26DTA and 34DTA, respectively. These interactions may lead to the transfer of charge from nitrogen to carbonyl group as well as thiazole ring. On the contrary, the second LP of O1, interacting with antibonds, *i.e.*, $n_2(O1) \to \sigma^*(C8\text{-}N1)$ and $n_2(O1) \to \sigma^*(C7\text{-}C8)$ provide a small stabilization. Note that the occupancy of O1 is larger than that of N1, which is almost equal to 2. In all molecules, primarily the p-character of oxygen LP orbitals along with the lower contribution in intra-molecular hyperconjugation exhibits features similar to pure LP orbitals. Thus the results of the NBO calculations suggest that oxygen atom is more reactive than nitrogen, possessing higher coordination ability.

Table 3. NBO analysis for lone-pair interaction in 26DTA, 24DTA, and 34DTA.

Molecule	Donor NBO (*i*)			Acceptor NBO (*j*)	$E^{(2)}$	E_j-E_i	F_{ij}
	Lone Pair	Occupancy	ED (%)		kcal/mol	a.u.	a.u.
26DTA	n_2(O1)			σ*(C1 - C7)	0.66	0.65	0.019
	n_2 (O1)			σ*(C7 - C8)	20.6	0.61	0.103
	n_2 (O1)			σ*(C8 - N1)	25.23	0.71	0.122
	n_2 (O1)	1.84817	99.5	σ*(S1 - C11)	2.16	0.49	0.030
	n_1 (N1)			π*(C8 - O1)	61.33	0.28	0.119
	n_1 (N1)	1.64542	99.9	π*(C9 - N2)	42.69	0.27	0.097
24DTA	n_2(O1)			σ*(C1 - C7)	0.60	0.66	0.018
	n_2(O1)			σ*(C7 - C8)	20.61	0.61	0.103
	n_2(O1)			σ*(C8 - N1)	25.4	0.71	0.122
	n_2(O1)	1.84997	99.5	σ*(S1 - C11)	2.15	0.49	0.03
	n_1(N1)			π*(C8 - O1)	57.71	0.29	0.117
	n_1(N1)	1.64614	99.9	π*(C9 - N2)	42.51	0.27	0.097
34DTA	n_2 (O1)			σ*(C1 - C7)	0.64	0.66	0.019
	n_2 (O1)			σ*(C7 - C8)	20.53	0.61	0.102
	n_2 (O1)			σ*(C8 - N1)	25.27	0.71	0.122
	n_2 (O1)	1.85022	99.5	σ*(S1 - C11)	2.16	0.49	0.030
	n_1 (N1)			π*(C8 - O1)	61.4	0.28	0.118
	n_1 (N1)	1.64368	99.9	π*(C9 - N2)	42.35	0.27	0.097

Now in order to analyze the relative coordination ability of molecules, we focus on the LP of O1 in 26DTA, 24DTA and 34DTA. In 24DTA, the net stabilization caused by different intra-molecular interactions is found to be 48.76 kcal/mol. The respective stabilization energies in 26DTA and 34DTA are 48.65 kcal/mol and 48.60 kcal/mol, respectively. Nevertheless, the occupancy in 34DTA is a bit higher than those in 26DTA and 24DTA. Hence, the smaller stabilization corresponding to hyperconjugations and greater occupancy of LP of O1 in 34DTA may indicate that this possesses relatively higher tendency of coordination, which is consistent with the absolute electronegativity discussed above. The charges on O1 atoms obtained by NPA (see Fig. 4) also accord this feature. One can see that the magnitude of the charge on O1, -0.612|e| for 34DTA is larger than -0.609|e| in 26DTA and 24DTA. It needs to be emphasized that there exists no significant difference in these molecules in terms of their coordination behavior.

SCALING OF NORMAL MODES OF VIBRATION

The analysis of normal modes is frequently used in the prediction and interpretation of the vibrational spectra of molecules (see Chapter 2). Nevertheless, there are inherent limitations of the approximations employed to obtain the frequencies of vibration. These are primarily due to the effect of electron–correlation (see Chapter 1), the deficiencies in the basis set and the role of inter- as well as intra-molecular interactions. One should always keep in mind that the vibrational frequencies are calculated for an isolated molecule in the gas phase, in contrast to experimental spectra determined in liquid or solid phase of sample with some impurities. It has been well established that $v_{gas} > v_{liquid} > v_{solid}$ [22]. The comparison of the experimental and calculated spectra indicates that if calculated frequencies are found to be smaller than the observed frequencies, there may exist weak inter-molecular interactions. This leads to the major drawback of vibrational spectroscopy, *i.e.*, the lack of a direct spectra–structure relation. DFT calculations use the model considering elastic chemical bonds, which becomes inappropriate particularly at higher frequencies due to anharmonicity in vibrations [23]. This anharmonicity affects several vibrations such as torsions and inversions, for instance, the umbrella modes of halomethyl radicals [24]. Likewise, for lower frequencies having rotational modes usually, the coupling of vibrational modes and molecular interaction come into effect, affecting the calculated frequencies. Therefore, the calculated frequencies by using the Born-Oppenheimer approximation (see Chapter 1) need to be properly scaled.

For the practical applications of vibrational spectroscopy, therefore, it is required to obtain calculated frequencies in close agreement with their observed values. This can generally be achieved by proper scaling of calculated frequencies. There is a lack of a universal scaling procedure, which is being paid attention by several researchers worldwide. Merrick *et al.* [25] suggested some scale factors on the basis of harmonic approximation, but they were not suitable, particularly for the region having mixing of several vibrations. According to this, the scaling factor for B3LYP/6-31+G(d,p) is:

$$v_{scaled} = 0.9648 \ v_{calculated} \qquad (1)$$

On the other hand, Alcolea Palafox *et al.* [26, 27] reported some scaling equations based on the statistical correlation and proposed that these equations provide better scaling as compared to constant scale factors. The scaling equation for B3LYP/6-31+G(d,p) is as follows:

$$\nu_{scaled} = 22.1 + 0.9543 \ \nu_{calculated} \tag{2}$$

Below we carry out a benchmark study regarding the scaling of vibrational wavenumbers (frequencies) of 26DTA by using two different schemes, eq. (1) and eq. (2). The calculated as well as scaled wavenumbers of normal modes of vibrations along with their assignments are listed in Table **4**. The FT-IR peaks of the significant modes are also given for comparison. To simplify the discussion on the performance of both scaling schemes for a given mode of vibration, we present the vibrational spectroscopic analysis into two parts as below:

Higher Wavenumber Region (Above 1400 cm^{-1})

In this region, the C-H stretching associated with the phenyl ring and methylene group, including N-H and C=O stretching of amide group are obtained. The N-H stretching mode is determined at 3203 cm^{-1} in the FT-IR, and their scaled wavenumbers are obtained at 3475 cm^{-1} and 3459 cm^{-1} by eq. (1) and eq. (2), respectively. The usual difference between experimental and theoretical wavenumbers is because of the existence of inter-molecular hydrogen bonding in the solid phase, which can not be taken into account for a single molecule. The most intense C-H vibration of the phenyl ring is observed at 3045 cm^{-1} in FT-IR, which corresponds to the scaled wavenumbers at 3090 cm^{-1} and 3078 cm^{-1} by eq. (1) and eq. (2), respectively.

The C=O stretching experimentally is determined at 1687 cm^{-1} and is scaled at 1684 and 1688 cm^{-1} by eq. (1) and eq. (2), respectively. Likewise, the C-C stretching corresponding to the phenyl ring at 1550 cm^{-1} has scaled wavenumbers of 1546 cm^{-1} and 1551 cm^{-1}. The C-H stretching of the methylene group mixed with the N-H stretching mode is observed at 1436 cm^{-1}, which corresponds to the theoretically scaled wavenumbers at 1422 cm^{-1} and 1428 cm^{-1} by eq. (1) and eq. (2), respectively.

Table 4. Normal modes assignments and scaled frequencies for 26DTA along with observed FT-IR bands [16].

Calculated Frequency (cm^{-1})	Scaled Frequency by eq. (1) (cm^{-1})	Scaled Frequency by eq. (2) (cm^{-1})	FTIR Value (cm^{-1})	Infrared Intensity (a.u.)	Assignment of Normal Modes of Vibration
3602	3475	3459	3203	68	ν(13N-14H)
3272	3156	3144	-	1	ν_s(15C-18H)+ ν_s(17C-21H)
3233	3119	3107	-	8	ν_{as}(15C-18H)+ ν_s(17C-21H)
3228	3114	3102	-	1.69	ν_{as}(3C-2H)+ν(2C-23H)+ ν(1C-7H)

(Table 4) cont.....

3203	3090	3078	3045	3	ν(3C-2H)+ν(2C-23H)+ ν(1C-7H)
3093	2984	2973	-	3	ν(9H-8C-10H)
1746	1684	1688	1687	266	ν (11C=12O)
1627	1569	1574	-	9	ν(C-C)R2
1603	1546	1551	1550	37	ν(C-C)R2
1582	1526	1531	-	430	ν(C-C)R2+ ν(C-N)R2+ ν(6C-13C)+ β(15-18H)
1525	1471	1477	-	107	β (15C-21H)+ β(15C-18H)+ β(13N-14H)
1474	1422	1428	1436	9	Scissoring (9H-8C-10H)+ β(13N-14H)
1468	1416	1423	-	63	β(C-H)R2)
1465	1413	1420	-	16	β(1C-7H)+ β(3C-22H)+ β(13N-14H)+ ν(5C-8C)
1459	1407	1414	-	26	Scissoring (9H-8C-10H)+ β(13N-14H)
1349	1301	1309	-	62	β(21C-7H)+ β(15C-18H)
1335	1288	1296	-	4	Rocking(9H-8C-10H)+ β(C-H)R1
1334	1287	1295	1286	1	τ(9H-8C-10H)+ β(C-H)R1
1306	1260	1268	-	112	β(13N-14H)+ ν(16C-19S)
1251	1206	1215	-	1	β(C-N)R2+τ (9H-8C-10H)
1226	1182	1192	-	53	β(1C-7H)+ β(3C-22H)+ β(15C-18H)+ Rocking(9H-8C-10H)
1220	1177	1186	-	43	β(15C-8H)+ β (13N-14H)+ β(3C-22H)+ Rocking(9H-8C-10H)
1196	1153	1163	-	9	β(C-H)R2+τ(9H-8C-10H)
1181	1139	1149	-	30	β(15C-18H)+ β(13N-14H)
1172	1130	1140	1141	16	τ (9H-8C-10H)+ β(2C-23H)
1101	1062	1072	-	13	β(C-H)R2
1088	1049	1060	-	9	Scissoring (18H-15C-21C-17H)
1086	1047	1058	-	8	β(3C-22H)+ β(1C-7H)+ Rocking(9H-8--10H)+ ν(6C-25Cl)
975	940	952	931	7	τ (9H-8C-10H)+ β(CH)R1
939	905	918	-	41	Rocking(9H-8C-10H)
897	865	878	-	3	τ(C-21 C-17H)
882	850	863	856	6	In plane R1bending
855	824	838	-	3	Ring R2 breathing (9H-8C-10H)
787	759	773	785	38	γ(C-H)R3
783	755	769	761	23	γ(C-H)R2+ R1 breathing
767	740	754	-	67	In plane R2 bending+ τ(9H-8C-10H)

(Table 4) cont.....

718	692	707	692	39	γ(15C-18H)+ γ(21C-17H)
661	637	652	-	11	β (R1)+R(9H-8C-10H)
631	608	624	-	29	γ(13N-14H)+ γ(16C-20N-15C)
624	602	617	592	8	γ(C-H)R1+ γ(19S-15C-20C)+Rocking(9H-8C-10H)
563	543	559	-	2	γ(R1)+ γ(13N-14H) +Rocking(9H-8C-10H)
559	539	555	-	12	Twisting R1+ Rocking(9H-8C-10H)
511	493	509	-	11	γ(R1)+ γ(13N-14H)
486	468	485	-	8	γ(1C-2C-3C)+ γ(CH)R2 + w(9H-8C-10H)
478	461	478	-	4	β(R2)+R(9H-8C-10H)
424	409	426	-	4	γ(4C-24Cl)+τ(1C-2C-3C)
404	389	407	-	5	β(CH)R2+R(9H-8C-10H)+ β(C-Cl)R2

Notes: ν – stretching; ν_s – symmetric stretching; ν_{as} – asymmetric stretching; β – in-plane-bending; γ – out-of-plane bending; ω – wagging; τ – torsion.

Lower Wavenumber Region (Below 1400 cm^{-1})

This region mainly includes the bending modes of the rings and various groups. The in-plane bending modes associated with the methylene group, determined at 1286 cm^{-1} and 1141 cm^{-1}, are found to be mixed with the phenyl ring vibration. The scale factor provides these modes at 1287 cm^{-1} and 1130 cm^{-1} whereas scaling equation gives at 1295 cm^{-1} and 1140 cm^{-1}, respectively. Likewise, the out-o--plane bending of methylene mixed with phenyl ring at 931 cm^{-1} in FT-IR is scaled at 940 cm^{-1} and 952 cm^{-1} by eq. (1) and eq. (2), respectively.

The in-plane bending of thiazole ring is determined at 856 cm^{-1}, corresponding to scaled wavenumbers of 850 cm^{-1} and 863 cm^{-1}. Furthermore, the bending associated with the phenyl ring measured at 785 cm^{-1} is found to be at 759 cm^{-1} and 773 cm^{-1} after scaling. Moreover, the mixing of vibrations of both rings is determined at 761 cm^{-1}, with respect to the scaled values of 755 cm^{-1} and 769 cm^{-1}. Even lower vibrational modes at 692 cm^{-1} and 592 cm^{-1} corresponding to the bending of thiazole rings are scaled at 692 cm^{-1} and 602 cm^{-1} by eq. (1), but at 707 cm^{-1} and 617 cm^{-1} by eq. (2).

Which is Better, Scale Factor or Scaling Equation?

One might be interested in knowing which scaling scheme should be preferred. In order to compare the performance of both schemes, we have displayed the statistical correlation between experimentally determined wavenumbers and theoretically scaled wavenumbers in Fig. (5). Both scaling schemes are capable of explain the FT-IR peaks, at least in the present case. We find the correlation coefficient of 0.9964 in either case, which ensures that the performance of both scaling schemes performs equally well.

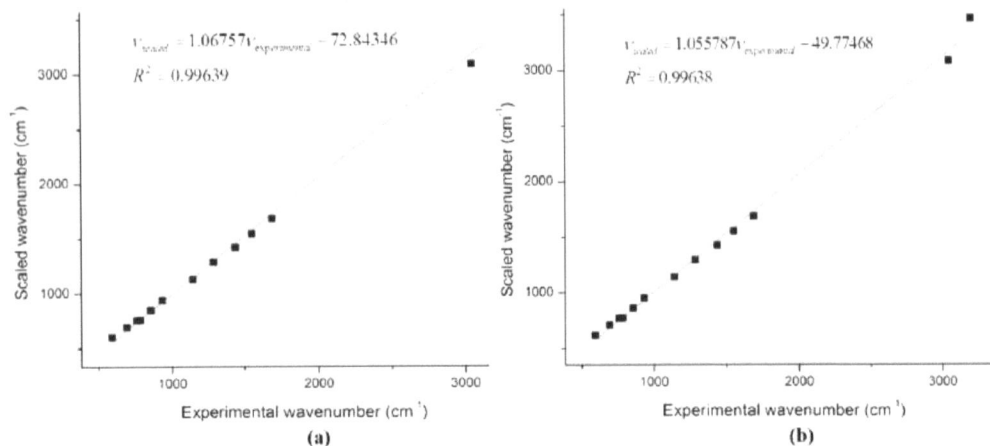

Fig. (5). Linear correlation between scaled and observed wavenumbers, (a) scale factor, eq. (1) and (b) scaling equation, eq. (2) [16].

The relative performance of both scaling schemes can be analyzed by calculating the difference of observed and scaled wavenumber as displayed in Fig. (6). The analysis of Fig. (6) suggests that the scaling equation works "slightly" better than scale factor particularly for the high wavenumber region. Therefore, the calculated wavenumbers are recommended to be scaled by a proper scaling equation to compensate for the anharmonicity of vibrations. In the case of the lower wavenumber region, where the difference between calculated and observed wavenumbers occurs primarily because of the coupling of vibrations and the effect of electron correlation, both scaling schemes works equivalently good.

As a practical consideration, the calculated wavenumbers are generally scaled by a scale factor as it is very easy to scale the wavenumbers by a constant factor than by any linear equation. That's why the vibrational modes discussed in succeeding chapters are scaled by an appropriate factor. However, below we compare the vibrational modes of 24DTA and 34DTA scaled by the scaling equation, eq. (2).

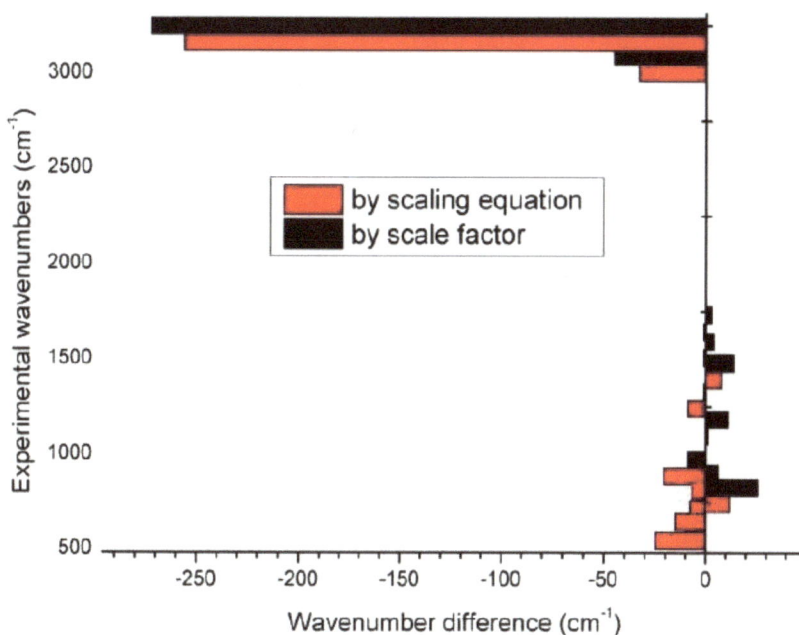

Fig. (6). Wavenumber difference by scaling schemes for 26DTA [16].

COMPARISON OF VIBRATIONAL MODES OF 24DTA AND 34DTA

All the vibrational modes of 24DTA and 34DTA were properly assigned on the basis of the potential energy distribution (PED) using the *VEDA 4* program [28, 29]. Tables **5** and **6** give calculated and scaled wavenumbers, FT-IR peaks along with relative intensity, IR intensities and vibrational assignments in 24DTA and 34DTA, respectively. The molecules consist of two ring systems in which a planar phenyl ring (R1) is linked to heterocyclic thiazole ring (R2) by acetamide fragment ($-NHCOCH_2-$).

Phenyl Ring (R1) Vibrations

These vibrations include the C-H stretching within the range 3100-3000 cm^{-1}, *i.e.*, the characteristic region in which the C-H stretching modes are identified [30]. The peaks present in this region remain almost independent of the nature of the substitutions. The calculated C-H stretchings are scaled in the region 3096-3060 cm^{-1}, and corresponding FT-IR peaks are determined at 3055 and 3034 cm^{-1} for 24DTA and 34DTA, respectively. These C-H stretching modes are pure with the PED contribution > 90%. Further, these vibrational modes are generally weak because of the transfer of charge from hydrogen to carbon as discussed earlier. The C-H bending, breathing and twisting modes are obtained below 1500 cm^{-1} with medium to weak intensities.

Table 5. Vibrational spectroscopic analysis of 24DTA [15].

Calculated Waven. (cm^{-1})	Scaled Waven. (cm^{-1})	FT-IR Peak (cm^{-1})	IR Intensity (a.u.)	Vibrational Assignments* [%PED]
3600	3458	3199 (s)	79	ν(N17–18H)[100]
3272	3145	-	1	ν(C–H)R2[99]
3243	3117	-	1	ν(C–H)R2[99]
3234	3108	-	8	ν$_{as}$(22H-19C-25C–21H)[89]
3226	3101	-	1	ν(C–H)R1[99]
3188	3064	3055 (w)	5	ν(C–H)R1[98]
3121	3001	-	1	ν$_{as}$(22H-19C-25C–21H)[89]
3068	2950	-	7	ν(13H-12C-14H)[91]
1748	1690	1691 (s)	273	ν(15C-16O)[83]
1632	1580	-	42	ν(C-C)R1[59]
1599	1548	1544 (m)	16	ν(C-C)R1[65]+ β (C–H)R1[14]
1581	1531	-	420	ν(20C-24N)[55]+ β (17N–18H)[21]
1524	1477	-	109	ν(19C-25C)[54]+ β (17N–18H)[12]+ β (C–H)R2[15]
1506	1459	-	84	β (C–H)R1[74]
1473	1428	-	6	Scissoring(13H-12C-14H[55])+ ν(20C-24N)[40]
1461	1416	-	38	Scissoring (13H-12C-14H)[72]+ β (17N–18H)[15]
1416	1373	1346 (m)	15	ν(C-C)R1[30]+ β (C-H)R1[27]
1348	1308	-	61	β (C-H)R1[86]
1339	1300	1300 (vw)	1	Rocking(13H-12C-14H)[39]+ β (C-H)R2[42]
1330	1291	-	3	Rocking(13H-12C-14H)[30]+ ν(C-C)R2[54]
1304	1267	-	107	ν(20C-24N)[45]+ β (17N–18H)[15]
1291	1254	1230 (w)	3	β (C-H)R1[76]
1227	1193	-	64	β (C-H)R1[54]+ β (17N-18H)[32]
1224	1190	-	3	β (C-H)R1[71]
1205	1172	-	31	Twist(13H-12C-14H)+ β (C-H)R2&R1
1178	1146	-	30	β (19C-22H)[65]+ β(17N-18H)[15]
1163	1132	-	4	β (1C-7H-6C-9H)[65]

(Table 5) cont.....

1119	1090	1103 (m)	50	β (3C-8H)[43]+ β (6C-9H)[32]
1088	1060	-	10	Scissoring (21H-25C-19C-22H)[56]
1066	1039	-	28	Breathing R2
976	954	960 (w)	7	Twist(13H-12C-14H)+S(16O-15C-17H)
967	945	-	1	Twist(1C-7H-6C-9H)
936	915	-	9	Rocking(13H-12C-14H)[45]
898	879	-	3	Twist(21C-25H-19C-22H)
885	867	866 (w)	11	γ(3C-8H)[49]
882	864	-	7	Breathing R2
871	853	-	43	Breathing R1
834	818	823 (m)	18	γ(1C-7H)[44]+ γ(6C-9H)[11]
786	772	771 (w)	7	γ(1C-7H)+ γ(6C-9H) + β(R1)
758	746	-	43	β (R2)
723	712	-	11	Ring R2 twist
717	706	698 (m)	43	γ(21C-25H)[38]+ γ(19C-22H)[12]
693	683	-	2	Ring R2 twist
677	668	-	2	T (13H-12C-14H)
648	641	-	2	γ(C-H)R1[54]
628	621	-	36	γ(C-H)R1+ γR2+Rocking(13H-12C-14H)
623	617	-	7	Twist R2+ Rocking(13H-12C-14H)
579	575	-	15	γ(7C-1H)[43]+ γ(3C-8H)[39]+ γ(2C-11Cl)[11]
564	560	-	4	Rocking(13H-12C-14H)[23]+ γ(19C-22H)[35]+ γ(17N-18H)[21]
524	522	-	16	R2 breathing + R1 twist
510	509	-	14	γ(17C-18H) + γ (ring R2)
486	486	-	8	Ring R2 twist+ γ(17C-18H)+ γ(23C-25C-20C)
452	453	-	3	γ(3C-8H)[35]+ γ(6C-9H)[22]+R(13H-12C-14H)[31]
437	439	-	3	Rocking(13H-12C-14H[29])+ γ(3C-8H)[35]+ γ(5C-9H)[12]
398	402	-	3	β (R1ring)
357	363	-	1	β (C-C)R2[23]+ β (15C-16O)[34]
321	328	-	1	γ(C-C)R2[26]+Rocking(13H-12C-14H)[17]
298	307	-	1	γ(23C-21H)+ γ(28C-17H)+Rocking(13H-12C-14H)
286	295	-	6	γ (Ring R2)

(Table 5) cont.....

267	277	-	4	β (11C-2Cl)[18]+ β (4C-10Cl)[13]+Rocking(13H-1-C-14H)[21]
201	214	-	1	β (11C-2Cl)[11]+ β (4C-10Cl)[27] +β (1C-7H)[23]
174	188	-	1	τ (Ring R1 &R2)
137	153	-	6	γ (Ring R1)
105	122	-	1	γ (Ring R2)
96	114	-	1	γ (Ring R2)
77	96	-	4	τ (13H-12C-14H)
32	53	-	1	Ring R1 & R2 Twist jointly
23	44	-	1	Ring R1 & R2 Twist jointly
19	40	-	1	Ring R1 & R2 Twist jointly

*Abbreviations: v – stretching; v_s – symmetric stretching; v_{as} – asymmetric stretching; β – in-plane-bending; γ – out-of-plane bending; ω – wagging; τ – torsion, s – scissoring. vs - very strong; s - strong; m - medium; w - weak; vw - very weak.

The C-C stretching modes are generally found in the region 1650-1200 cm^{-1} [31]. These vibrations are very sensitive, which depend on the nature of the substitution. These modes are scaled at 1580 cm^{-1}, 1548 cm^{-1} as well as 1477 cm^{-1} in 24DTA and 1583 cm^{-1}, 1549 cm^{-1} as well as 1476 cm^{-1} in 34DTA. The C-C bending, twisting and breathing modes, coupled with other modes are obtained in the lower wavenumber region. The C-C modes are relatively more intense than C-H modes.

Thiazol Ring (R2) Vibrations

The C-H stretching vibrations corresponding to the thiazole ring are scaled in the region 3145-2950 cm^{-1} in both compounds. These modes are not sensitive to the substitutions in the ring, which are in accordance with the reported peaks in thiazoles and related compounds [32]. The strongest peaks associated with the C-N stretching of thiazole are scaled in the range 1532-1531 cm^{-1}. These normal modes of vibrations are mixed with the N-H stretching, which are experimentally determined in the region 1544-1527 cm^{-1} by the FT-IR spectra. Some weak modes associated with the thiazole ring corresponding to breathing and twisting vibrations are obtained below 1500 cm^{-1} but close to 600 cm^{-1} in the case of the modes incorporating the vibrations of S atom.

Table 6. Vibrational spectroscopic analysis of 34DTA [15].

Unscaled Waven. (cm⁻¹)	Scaled Waven. (cm⁻¹)	FTIR Data (Intensity) (cm⁻¹)	IR Intensity (a.u.)	Mode of Vibration [%PED]
3598	3456	3197 (s)	97	ν(15N–16H)[100]
3271	3144	-	1	ν(23C–19H)[44]+ ν(17C–20H)[45]
3234	3108	-	8	ν(23C–19H)[99]
3220	3095	-	1	ν(1C–7H)+ ν(6C–8H)
3204	3080	-	1	ν(9C–23H)[99]
3191	3067	3034 (w)	4	ν(6C–8H)[48]+ ν(1C–7H)[55]
3106	2986	-	1	ν$_{as}$(12H-110C-11H)[89]
3065	2947	-	8	ν(12H-10C-11H)[91]
1743	1685	1689 (vs)	289	ν(15C-16O)[83]
1636	1583	1589 (w)	5	ν(C-C)R1[59]
1600	1549	-	13	ν(C-C)R1[65]+ β (C–H)R1[14]
1582	1532	1527 (vs)	461	ν(18C-22N)[55]+ β (15N–16H)[21]
1524	1476	-	104	ν(C-C)R2[54]+ β (15N–16H)[12]+ β (17C–20H)[15]
1506	1459	1462 (s)	74	β (C–H)R2[74]+ ν(C-C)R2[21]
1476	1431	-	3	Scissoring(13H-12C-14H[55])+ ν(C-N)R2[40]
1461	1416	-	38	Scissoring (13H-12C-14H)[72]+ β (17N–18H)[15]
1425	1382	-	41	ν(C-C)R1[30]+ β (C–H)R1[27]
1349	1310	-	59	β (C-H)R1[46]+ ν(C-C)R1[30]
1331	1292	1294 (w)	9	Rocking(13H-12C-14H)[39]+ β (C–H)R1[42]
1328	1289	-	1	Rocking(13H-12C-14H)[39]+ν(C-C)R1[32]
1304	1267	-	107	β (15C-16H)[26]+ ν(C-C)R1[54]
1285	1248	1228 (w)	3	β (C-H)R1[76]
1228	1194	-	15	β (17C-20H) [19]+ Rocking (11H-10C-12H)[22]+ β (C-H)R1[26]
1222	1188	-	3	β (17C- 20H) [19]+ β (15N-16H)[22]+ β (4C-24H)R1[26]
1204	1171	-	2	Twist(11H-10C-12H)+ β (6C-8H)
1180	1148	-	35	β (17C-18H)R[65]+ β(17N-18H)[15]
1168	1137	-	3	β (1C-7H)[65]+ β (6C-8H)[14]
1148	1118	-	50	β (C-C)R1[43]+ β (C-H)R1[32]
1089	1061	-	10	β (17C-20H) [43]+ β (23C-10H)[32]
1043	1017	-	35	β (C- H)R1[28]+ β (C- C)R1
977	955	-	7	Twist(13H-12C-14H)+Scissoring(16O-15C-17H)

(Table 6) cont.....

970	948	-	1	Twist(7C-1H-6C-8H)[59]
943	922	-	12	Rocking(13H-12C-14H)[45]+ γ(4C-24H)[29]
909	890	-	11	γ(C-H)[23]+ γ(C-C)[49]
898	879	-	3	γ(C-H)R2[49]
891	872	-	10	Breathing R1+ γ(4C-24H)
881	863	-	7	Breathing R2
840	824	815 (m)	17	γ(6C-8H)[44]+ γ(1C-7H)[11]
781	767	763 (m)	23	γ(6C-8H)+ β(Ring R1)
737	725	729 (w)	8	γ(C-C)R1[44]+ γ(C-H)R1[11]
719	708	-	43	γ (23C-14H)[32]+ γ(17C-20H)[53]
708	698	-	11	γ(C-C)[38]+ Rocking(11-10C-12H)[12]
699	689	-	3	γ(C-H)[23]+ γ(C-C)[49]
681	672	-	10	τ (13H-12C-14H)
652	644	-	17	γ(C-S-C)R2[54]
634	627	-	27	γ(15C-16H)R2+ γR2+Rocking(11H-10C-12H)
624	618	-	17	γ(C-S-C)R2 [43]+ γ(C-H)R2[39]
598	593	580 (w)	9	Twist R1+ Rocking(13H-12C-14H)
566	562	-	5	γ(R2)+ γ(15N-16H)
526	524	-	21	R1 breathing + R2 twist
510	509	-	14	γ(17C-18H) + ring R2 out of plane bending
483	483	-	6	γ(CCC)R1+ γ(C-Cl)R1
460	461	-	1	Ring R1 twist
448	450	-	4	γ(CCC)R1[34]+ γ(1C-7H)[54]
386	391	-	1	τ (Ring R1)
356	362	-	1	β (C-C)R1[23]+ β (15N-16H)[34]
319	327	-	3	γ(C-C)R1[26]+ β(13C-14)[25])
295	304	-	1	γ (Ring R2)
288	297	-	3	γ(CCC)R1+ γ(C-H)R1
243	254	-	1	τ(11H-10C-12H)
199	212	-	1	β (C-Cl)R1[11+β (1C-7H)[23]
183	197	-	2	τ (Ring R1)
135	151	-	5	τ(CCC)+ γ(CH)R1
105	122	-	1	τ (Ring R1)
94	112	-	0	τ (Ring R2)
76	95	-	6	τ(11H-10C-12H)+ γ(13C-14H)

(Table 6) cont.....

33	54	-	1	τ (Ring R1)
16	37	-	1	γ(R1)
12	34	-	1	Ring R1 & R2 twist jointly

Fragment (-NHCOCH₂-) Vibrations

The broad and intense bands associated with amides are found in the region 3500-3100 cm⁻¹ for N-H stretching and 1700-1650 cm⁻¹ for C=O stretching [22]. The C=O vibrations are very delicate due to the fact both carbon and oxygen atoms are displaced with almost equal amplitude of the vibration. The values scaled within the range 3458-3456 cm⁻¹ for N-H stretching and 1690-1685 cm⁻¹ for C=O stretching are in good agreement with the values reported in literature. From FT-IR spectra, the respective C=O stretching modes are measured at 1691-1689 cm⁻¹ while N-H stretching modes are achieved in the lower range at 3199-3197 cm⁻¹. The apparent discrepancy between experimentally observed and theoretically scaled N-H stretching can be expected due to the existence of inter-molecular H-bonds in condensed phase.

The methylene (-CH₂) group stretching associated with the acetamide fragment and corresponding bending (scissoring and rocking) are essentially found in the region 3000-2850 cm⁻¹ and below 1500 cm⁻¹, respectively [32]. These modes are quite obvious, which are of no more practical importance. The -CH₂ stretching vibration corresponds to the scaled wavenumber at 2950 cm⁻¹ and 2947 cm⁻¹ in 24DTA and 34DTA, respectively.

CONCLUDING REMARKS

We discussed the results of DFT calculations on three acetamides, 26DTA, 24DTA, and 34DTA. The calculated structural parameters are in good agreement with crystallographic data. Vibrational frequencies were scaled by two different schemes, eq. (1) and (2). In general, both schemes efficiently reproduce observed spectrum in the case of 26DTA. In the high frequency region, the scaling equation, eq. (2) performs better than scale factor, eq. (1). For 24DTA and 34DTA also, the scaled theoretical wavenumbers are in perfect agreement with the experimental FTIR values and the vibrational modes are interpreted in terms of the PED contribution. The NBO analysis, in conjunction with NPA, revealed potential coordinating ability of molecules due to mainly p character, larger occupancy and less engagement in intramolecular hyperconjugative interaction of O lone pair. Electronic parameters values indicate variation in chemical reactivity and polarity of molecules with the position of substitution. The intramolecular interaction in these molecules revealed by the QTAIM will be discussed in Chapter 7.

CONSENT FOR PUBLICATION

Not applicable.

CONFLICT OF INTEREST

The authors declare no conflict of interest, financial or otherwise.

ACKNOWLEDGEMENTS

Declared none.

REFERENCES

[1]   Cantello, B.C.C.; Cawthorne, M.A.; Cottam, G.P.; Duff, P.T.; Haigh, D.; Hindley, R.M.; Lister, C.A.; Smith, S.A.; Thurlby, P.L. [[omega-(Heterocyclylamino)alkoxy]benzyl]-2,4-thiazolidinediones as potent antihyperglycemic agents. *J. Med. Chem.,* **1994**, *37*(23), 3977-3985.
        [http://dx.doi.org/10.1021/jm00049a017] [PMID: 7966158]

[2]   Küçükgüzel, G.; Kocatepe, A.; De Clercq, E.; Sahin, F.; Güllüce, M. Synthesis and biological activity of 4-thiazolidinones, thiosemicarbazides derived from diflunisal hydrazide. *Eur. J. Med. Chem.,* **2006**, *41*(3), 353-359.
        [http://dx.doi.org/10.1016/j.ejmech.2005.11.005] [PMID: 16414150]

[3]   Quiroga, J.; Hernandez, P.; Insuassty, B.R. Control of the reaction between 2-aminobenzothiazoles and Mannich bases. Synthesis of pyrido[2,1-b][1,3]benzothiazoles versus [1,3]benzothiazolo[2,3-b]quinazolines, J. Chem. Soc. *Perkin Trans.,* **2002**, *1*, 555-559.
        [http://dx.doi.org/10.1039/b109676a]

[4]   Hutchinson, I.; Jennings, S.A.; Vishnuvajjala, B.R.; Westwell, A.D.; Stevens, M.F. Antitumor benzothiazoles. 16. Synthesis and pharmaceutical properties of antitumor 2-(--aminophenyl)benzothiazole amino acid prodrugs. *J. Med. Chem.,* **2002**, *45*(3), 744-747.
        [http://dx.doi.org/10.1021/jm011025r] [PMID: 11806726]

[5]   Hargrave, K.D.; Hess, F.K.; Oliver, J.T.N. N-(4-substituted-thiazolyl)oxamic acid derivatives, a new series of potent, orally active antiallergy agents. *J. Med. Chem.,* **1983**, *26*(8), 1158-1163.
        [http://dx.doi.org/10.1021/jm00362a014] [PMID: 6876084]

[6]   Patt, W.C.; Hamilton, H.W.; Taylor, M.D.; Ryan, M.J.; Taylor, D.G., Jr; Connolly, C.J.; Doherty, A.M.; Klutchko, S.R.; Sircar, I.; Steinbaugh, B.A. Structure-activity relationships of a series of 2-amino-4-thiazole-containing renin inhibitors. *J. Med. Chem.,* **1992**, *35*(14), 2562-2572.
        [http://dx.doi.org/10.1021/jm00092a006] [PMID: 1635057]

[7]   Sharma, P.K.; Sawhney, S.N.; Gupta, A.; Singh, G.B.; Bani, S. Synthesis and Antiinflammatory Activity of Some 3-(2-Thiazolyl)-1, 2-benzisothiazoles. *Indian J. Chem. B,* **1998**, *37*, 376-381.

[8]   Jaen, J.C.; Wise, L.D.; Caprathe, B.W.; Tecle, H.; Bergmeier, S.; Humblet, C.C.; Heffner, T.G.; Meltzer, L.T.; Pugsley, T.A. 4-(1,2,5,6-Tetrahydro-1-alkyl-3-pyridinyl)-2-thiazolamines: a novel class of compounds with central dopamine agonist properties. *J. Med. Chem.,* **1990**, *33*(1), 311-317.
        [http://dx.doi.org/10.1021/jm00163a051] [PMID: 1967314]

[9]   Tsuji, K.; Ishikawa, H. Synthesis and anti-pseudomonal activity of new 2-isocephems with a dihydroxypyridone moiety at C-7. *Bioorg. Med. Chem. Lett.,* **1994**, *4*, 1601-1606.
        [http://dx.doi.org/10.1016/S0960-894X(01)80574-6]

[10]   Bell, F.W.; Cantrell, A.S.; Högberg, M.; Jaskunas, S.R.; Johansson, N.G.; Jordan, C.L.; Kinnick, M.D.; Lind, P.; Morin, J.M., Jr; Noréen, R. Phenethylthiazolethiourea (PETT) compounds, a new class of HIV-1 reverse transcriptase inhibitors. 1. Synthesis and basic structure-activity relationship studies

of PETT analogs. *J. Med. Chem.,* **1995**, *38*(25), 4929-4936.
[http://dx.doi.org/10.1021/jm00025a010] [PMID: 8523406]

[11]   Ergenç, N.; Çapan, G.; Günay, N.S.; Özkirimli, S.; Güngör, M.; Özbey, S.; Kendi, E. Synthesis and hypnotic activity of new 4-thiazolidinone and 2-thioxo-4,5-imidazolidinedione derivatives. *Arch. Pharm. (Weinheim),* **1999**, *332*(10), 343-347.
[http://dx.doi.org/10.1002/(SICI)1521-4184(199910)332:10<343::AID-ARDP343>3.0.CO;2-0]
[PMID: 10575366]

[12]   Carter, J.S.; Kramer, S.; Talley, J.J.; Penning, T.; Collins, P.; Graneto, M.J.; Seibert, K.; Koboldt, C.M.; Masferrer, J.; Zweifel, B. Synthesis and activity of sulfonamide-substituted 4,5-diaryl thiazoles as selective cyclooxygenase-2 inhibitors. *Bioorg. Med. Chem. Lett.,* **1999**, *9*(8), 1171-1174.
[http://dx.doi.org/10.1016/S0960-894X(99)00157-2] [PMID: 10328307]

[13]   Badorc, A.; Bordes, M.F.; de Cointet, P.; Savi, P.; Bernat, A.; Lalé, A.; Petitou, M.; Maffrand, J.P.; Herbert, J.M. New orally active non-peptide fibrinogen receptor (GpIIb-IIIa) antagonists: identification of ethyl 3-[N-[4-[4-[amino[(ethoxycarbonyl) imino]methyl]phenyl]-1,3-thiazol-2--l]-N-[1-[(ethoxycarbonyl)methyl]pip erid -4-yl]amino]propionate (SR 121787) as a potent and long-acting antithrombotic agent. *J. Med. Chem.,* **1997**, *40*(21), 3393-3401.
[http://dx.doi.org/10.1021/jm970240y] [PMID: 9341914]

[14]   Rudolph, J.; Theis, H.; Hanke, R.; Endermann, R.; Johannsen, L.; Geschke, F. seco-Cyclothialidines: new concise synthesis, inhibitory activity toward bacterial and human DNA topoisomerases, and antibacterial properties. *J. Med. Chem.,* **2001**, *44*(4), 619-626.
[http://dx.doi.org/10.1021/jm0010623] [PMID: 11170652]

[15]   Srivastava, A.K.; Pandey, A.K.; Narayana, B.; Sarojini, B.K.; Nayak, P.S.; Misra, N. Normal Modes, Molecular Orbitals and Thermochemical Analyses of 2,4 and 3,4 Dichloro Substituted Phenyl-N-(1-3-thiazol-2-yl)acetamides: DFT Study and FTIR Spectra. *J. Theor. Chem.,* **2014**, *10*, 125841.
[http://dx.doi.org/10.1155/2014/125841]

[16]   Srivastava, A.K.; Pandey, A.K.; Pandey, S.; Nayak, P.S.; Narayana, B.; Sarojini, B.K.; Misra, N. Uniform versus Nonuniform Scaling of Normal Modes Predicted by Ab Initio Calculations: A Test on 2-(2,6-Dichlorophenyl)-N-(1,3-thiazol-2yl) Acetamide. *Int. J. Spectrosc.,* **2014**, *7*, 649268.
[http://dx.doi.org/10.1155/2014/649268]

[17]   Srivastava, A.K.; Misra, N. A comparative theoretical study on the biological activity, chemical reactivity, and coordination ability of dichloro-substituted (1,3-thiazol-2-yl)acetamides. *Can. J. Chem.,* **2014**, *92*, 234-239.
[http://dx.doi.org/10.1139/cjc-2013-0335]

[18]   Nayak, P.S.; Narayana, B.; Yathirajan, H.S.; Jasinski, J.P.; Butcher, R.J.; Butcher, R.J. 2-(2,--Dichloro-phen-yl)-N-(1,3-thia-zol-2-yl)acetamide. *Acta Crystallogr. Sect. E Struct. Rep. Online,* **2013**, *69*(Pt 4), o523.
[http://dx.doi.org/10.1107/S1600536813006260] [PMID: 23634066]

[19]   Nayak, P.S.; Narayana, B.; Yathirajan, H.S.; Jasinski, J.P.; Butcher, R.J.; Butcher, R.J. 2-(2,4--i-chloro-phen-yl)-N-(1,3-thia-zol-2-yl)acetamide. *Acta Crystallogr. Sect. E Struct. Rep. Online,* **2013**, *69*(Pt 5), o656-o657.
[http://dx.doi.org/10.1107/S1600536813008532] [PMID: 23723819]

[20]   Nayak, P.S.; Narayana, B.; Yathirajan, H.S.; Jasinski, J.P.; Butcher, R.J.; Butcher, R.J. 2-(3,--Dichloro-phen-yl)-N-(1,3-thia-zol-2-yl)acetamide. *Acta Crystallogr. Sect. E Struct. Rep. Online,* **2013**, *69*(Pt 5), o645-o646.
[http://dx.doi.org/10.1107/S1600536813008374] [PMID: 23723810]

[21]   Wu, W.N.; Cheng, F.X.; Yan, L.; Tang, N. Synthesis, characterization and fluorescent properties of lanthanide complexes with two aryl amide ligands. *J. Coord. Chem.,* **2008**, *61*, 2207-2215.
[http://dx.doi.org/10.1080/00958970801901329]

[22]   Silverstein, M.; Basseler, G.C.; Morill, C. *Spectrometric Identification of Organic Compounds*; Wiley:

New York, USA, **1981**.

[23]   Pople, J.A.; Scott, A.P.; Wong, M.W.; Radom, L. Scaling factors for obtaining fundamental vibrational frequencies and zero-point energies from HF/6–31G* and MP2/6–31G* harmonic frequencies. *Isr. J. Chem.,* **1993**, *33*, 345-350.
       [http://dx.doi.org/10.1002/ijch.199300041]

[24]   Marshall, P.; Srinivas, G.N.; Schwartz, M. A computational study of the thermochemistry of bromine- and iodine-containing methanes and methyl radicals. *J. Phys. Chem. A,* **2005**, *109*(28), 6371-6379.
       [http://dx.doi.org/10.1021/jp0518052] [PMID: 16833980]

[25]   Alecu, I.M.; Zheng, J.; Zhao, Y.; Truhlar, D.G. Computational thermochemistry: scale factor databases and scale factors for vibrational frequencies obtained from electronic model chemistries. *J. Chem. Theory Comput.,* **2010**, *6*(9), 2872-2887.
       [http://dx.doi.org/10.1021/ct100326h] [PMID: 26616087]

[26]   Alcolea Palafox, M. Scaling factors for the prediction of vibrational spectra. I. Benzene molecule. *Int. J. Quantum Chem.,* **2000**, *77*, 661-684.
       [http://dx.doi.org/10.1002/(SICI)1097-461X(2000)77:3<661::AID-QUA7>3.0.CO;2-J]

[27]   Alcolea Palafox, M.; Nunez, J.L.; Gil, M. Accurate scaling of the vibrational spectra of aniline and several derivatives. *J. Mol. Struct. THEOCHEM,* **2002**, *593*, 101-131.
       [http://dx.doi.org/10.1016/S0166-1280(02)00319-6]

[28]   Jamroz, M. H. *Vibrational energy distribution analysis VEDA 4,* **2004**.

[29]   Jamróz, M.H. Vibrational energy distribution analysis (VEDA): scopes and limitations. *Spectrochim. Acta A Mol. Biomol. Spectrosc.,* **2013**, *114*, 220-230.
       [http://dx.doi.org/10.1016/j.saa.2013.05.096] [PMID: 23778167]

[30]   Rastogi, V.K.; Alcolea Palafox, M.; Tanwar, R.P.; Mittal, L. 3,5-Difluorobenzonitrile: ab initio calculations, FTIR and Raman spectra. *Spectrochim. Acta A Mol. Biomol. Spectrosc.,* **2002**, *58*(9), 1987-2004.
       [http://dx.doi.org/10.1016/S1386-1425(01)00650-3] [PMID: 12164497]

[31]   Barnes, A.J.; Majid, M.A.; Stuckey, M.A.; Gregory, P.; Stead, C.V. The resonance Raman spectra of Orange II and Para Red: molecular structure and vibrational assignment. *Spectrochim. Acta A,* **1985**, *41*, 629-635.
       [http://dx.doi.org/10.1016/0584-8539(85)80050-7]

[32]   Taurins, A.; Fenyes, J.G.E.; Jones, R.N. Thiazoles: iii. Infrared spectra of methylthiazoles. *Can. J. Chem.,* **1957**, *35*, 423-427.
       [http://dx.doi.org/10.1139/v57-061]

CHAPTER 4

DFT Study on an Unnatural Amino Acid: 4-Hydroxyproline

Abstract: 4-hydroxy-*l*-proline is formed by hydroxylation of proline, an amino acid found in protein, whose inhibition results in hair problems in humans, causing scurvy disease. In this chapter, we discuss the DFT study on *cis* and *trans* conformers of 4-hydroxy-*l*-proline, *i.e.*, CHLP and THLP using the B3LYP/6-31+G(d,p)level. The equilibrium structures of both conformers are obtained to analyze their vibrational properties. We have also discussed the results of an in-depth study on *cis*-4-hydroxy-*d*-proline (CHDP). The scan of potential energy surface provides the global minimum structure of CHDP along with its possible conformers. HOMO, LUMO, and MESP surfaces as well as charge distribution, are used to explain the chemical reactivity. The equilibrium geometry of CHDP dimer has been obtained and intermolecular interactions are explored by the NBO analyses. Vibrational analysis has been carried out on CHLP, THLP, and CHDP (monomer and dimer). A complete assignment to vibrational modes has been presented based on their potential energy distribution. The calculated frequencies, after proper scaling, are compared with experimental FT-IR frequencies recorded by KBr disc and Nujol mull techniques. Several electronic as well as thermodynamic parameters have also been reported.

Keywords: B3LYP, Charge distribution, DFT, Dimer, Electronic parameter, FT-IR, H-bond, HOMO-LUMO, Hydroxyproline, Inter-molecular interaction, MESP, NBO, Proline, Thermodynamic parameter, Vibrational spectra.

INTRODUCTION

The protein structure and its functions have been the subject of investigation for a long time [1]. Proteins contain many amino acids of different types with complex structures. One of them is proline, which is unique due to the fact that the amine nitrogen is attached to two alkyl groups. Consequently leading it to a secondary amine. Proline possesses an enormous conformational rigidity due to the distinguishable cyclic form of its side-chain. This side-chain also changes the rate of the formation of peptide bonds among proline and other amino acids. The incorporation of unnatural amino acids leads to the exploration of protein structure and functions [2, 3]. Proline and related compounds are frequently employed in organic reactions as asymmetric catalysts. Furthermore, proteins with

Ambrish Kumar Srivastava and Neeraj Misra

excess proline interact with polyphenols to form haze (turbidity) during brewing [4, 5].

4-Hydroxyproline is derived from proline by substituting a hydroxyl ($-$OH) to the gamma carbon atom. Despite the fact that it is not directly found into proteins, it consists of almost 4% of all amino acids obtained in animal tissue, which exceeds several translationally incorporated amino acids [6]. Hydroxyproline becomes a main constituent of the protein collagen [7], playing an important role in collagen stability [8] by facilitating the sharp twisting of the collagen helix [9]. Therefore, it can be exploited as an indicator to estimate the amount of collagen and/or gelatin. The glycoproteins incorporating hydroxyproline have also been obtained in plant cell walls [10]. The hydroxylation of proline needs ascorbic acid, whose absence in humans results in the reduced stability of the collagen due to defect in the hydroxylation of proline residues of collagen, leading to scurvy disease. Nevertheless, the increased serum and urine levels of hydroxyproline are obtained in the case of Paget's disease [11]. 4-hydroxyproline is also found in the toxic cyclic peptides from Amanita mushrooms (*e.g.* alpha-amanitin and phalloidin) [12].

Many chemical and/or biological properties are closely associated with the molecular structures or geometries. The different conformations of the same molecule may lead to different chemical properties. The aim of the present study includes detailed structural and vibrational analyses of two potential conformers of 4-hydroxyproline, namely, *cis*-4-hydroxy-*l*-proline (CHLP) and *trans*-4-hydroxy-*l*-proline (THLP). In this chapter, we will first provide a comparative analysis of CHLP and THLP [13]. Subsequently, we will focus on the *cis*-4-hydroxy-*d*-proline (CHDP) and its dimer [14]. The CHLP and CHDP can be differentiated on the basis of the circular dichroism or optical rotatory dispersion, which will not be discussed here. All results described in this chapter are obtained at the B3LYP/6-31+G** or B3LYP/6-31+G(d,p) level using the *Gaussian 09* program.

CIS-4-HYDROXY-*L*-PROLINE AND *TRANS*-4-HYDROXY-*L*-PROLINE

Structural Properties

The optimized geometries of CHLP and THLP are shown in Fig. (**1**) and respective bond-lengths can be found in Table **1**. 4-hydroxyproline is composed of one hetero-pentagonal ring with one carbon replaced by nitrogen. This pentagonal ring deviates from planarity because of the repulsion due to non-bonding electrons of nitrogen. The covalent bond between the side-chain and nitrogen backbone has relevant structural consequences on the properties of both conformers. However, there is a strict dependence between the main- and side-chain proline conformations. This is reflected in the non-bonding distances between backbone atoms, R(1–4) which is 2.12 Å for CHLP and 2.118 Å for THLP and R(1–5), 2.128 Å (CHLP) and 2.12 Å (THLP) (not given in Table **1**).

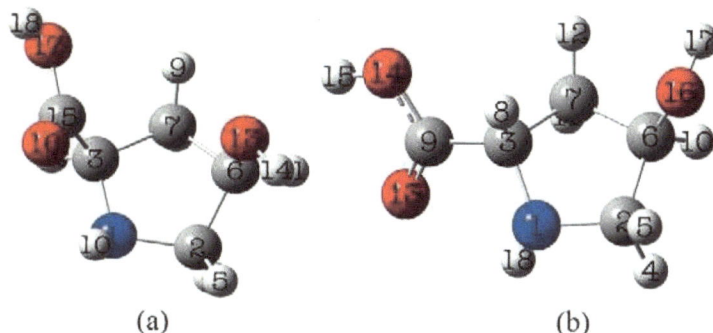

(a)   (b)

Fig. (1). Optimized geometries of **(a)** CHLP and **(b)** THLP [13].

In *cis* and *trans* conformers, the same groups (−OH and −COOH) are present at different angular positions of the hetero-pentagonal ring. This results in the change in ring geometries which have also been proven to be a sensitive indicator of the interaction between the substituent and the ring. Consequently, a re-adjustment takes place in the ring due to the presence of these groups in both *cis* and *trans* conformations. The average bond-length of C−C in the ring for CHLP (1.52 Å) takes a slightly lower value than in THLP (1.53 Å). However, C−N bond-length in *cis* (1.47 Å) becomes slightly higher than that in *trans* conformer, 1.46 Å (see Table **1**). In general, calculated bond-lengths are consistent with the observed values for 4-hydroxyproline [15] by X-ray and neutron diffraction. For instance, the C−C bond-lengths in 4-hydroxyproline are observed to be between 1.52 Å and 1.53 Å.

Table 1. Bond-lengths (R, in Å) of CHLP and THLP calculated at the B3LYP/6-31+G(d,p) level. Refer to (Fig. 1) for atomic labeling [13].

CHLP				THLP			
Parameter	**Value**	**Parameter**	**Value**	**Parameter**	**Value**	**Parameter**	**Value**
R(1-2)	1.475	R(6-7)	1.528	R(1-2)	1.475	R(6-7)	1.539
R(1-3)	1.469	R(6-11)	1.099	R(1-3)	1.462	R(6-10)	1.098
R(1-10)	1.016	R(6-13)	1.432	R(1-18)	1.017	R(6-16)	1.434
R(2-4)	1.099	R(7-8)	1.094	R(2-4)	1.094	R(7-11)	1.097
R(2-5)	1.096	R(7-9)	1.091	R(2-5)	1.095	R(7-12)	1.094
R(2-6)	1.537	R(13-14)	0.966	R(2-6)	1.553	R(9-13)	1.214

(Table 1) cont.....

CHLP				THLP			
R(3-7)	1.555	R(15-16)	1.213	R(3-7)	1.555	R(9-14)	1.354
R(3-12)	1.099	R(15-17)	1.354	R(3-8)	1.096	R(14-15)	0.973
R(3-15)	1.525	R(17-18)	0.972	R(3-9)	1.512	R(16-17)	0.966

Vibrational Properties

In this section, we discuss the vibrational properties of CHLP and THLP. The calculated frequencies are scaled by a scale-factor [16] of 0.9648 (as mentioned in Chapter 2) and compared with corresponding FTIR values. The FTIR spectra of both conformers are obtained from the SDBS website [17] as shown in Fig. (2). In Fig. (3), we have shown a correlation between calculated and experimental frequencies. In both cases, the correlation coefficients (R^2) are close enough to unity, ensuring the reliability of our computations.

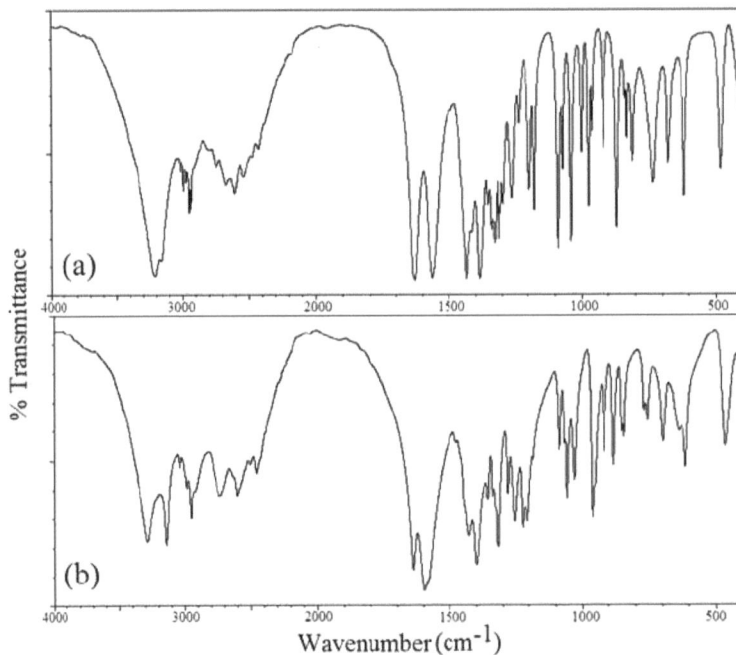

Fig. (2). FTIR spectra of **(a)** CHLP and **(b)** THLP [13].

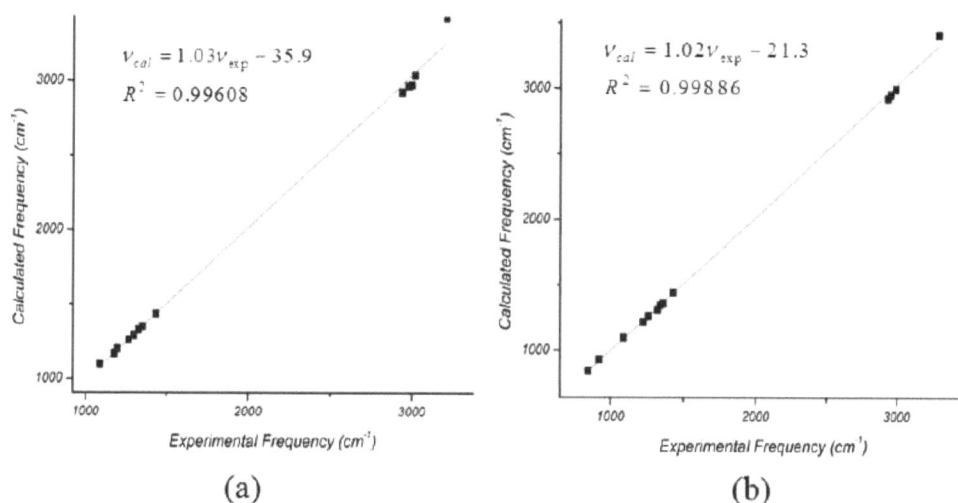

$$v_{cal} = 1.03 v_{exp} - 35.9$$
$$R^2 = 0.99608$$

$$v_{cal} = 1.02 v_{exp} - 21.3$$
$$R^2 = 0.99886$$

(a)   (b)

Fig. (3). Correlation between calculated (scaled) frequency (v_{cal}) and experimental frequency (v_{exp}) for **(a)** CHLP and **(b)** THLP [13].

Table 2. Proposed assignments and potential energy distribution (PED) for vibrational modes of CHLP calculated at the B3LYP/6-31+G(d,p) level [13].

Calc. Freq. (cm⁻¹)	Scaled Freq. (cm⁻¹)	IR Int. (a.u.)	FTIR Freq. (cm⁻¹)	Assignments (PED ≥10%)
3826	3691	26.3	-	v(O13H14)(100)
3757	3624	67.5	-	v(O17H18)(100)
3546	3421	31.8	3212	v(N1H10)(100)
3150	3039	5.0	3016	v(C7H9)(84)+ v(C7H8)(16)
3082	2973	32.7	2996	v(C7H8)(82)+ v(C7H9)(15)
3075	2966	31.6	2976	v(C2H5)(90)
3032	2925	88.7	2938	v(C6H11)(84)+ v(C2H4)(12)
3018	2911	27.8	-	v(C3H12)(97)
3010	2904	35.8	-	v(C2H4)(77)+ v(C6H11)(13)
1810	1746	296	-	v(C15O16)(78)
1509	1455	0.5	-	ρ(CH₂)(85)+ρ(CH₂)(11)
1489	1436	7.1	1434	ρ(CH₂)(84)+ρ(CH₂)(12)
1481	1428	12.0	-	δ(C2H10N1)(71)
1400	1350	4.3	1353	δ(H11O13C6)(36)+ δ(C6H14O13)(12)+ v(C6C7)(10)
1378	1329	22.5	1328	ω(N1H4C2)(23)+ δ(H11O13C6)(12)+ω(C6H8C7)(10)
1371	1322	38.5	-	ω(N1H4C2)(21)+ρ(CH₂)(12)

(Table 2) cont.....

Calc. Freq. (cm⁻¹)	Scaled Freq. (cm⁻¹)	IR Int. (a.u.)	FTIR Freq. (cm⁻¹)	Assignments (PED ≥10%)
1339	1291	25.8	1298	ω(C2H11C6)(29)+ ν(C2C6)(12)+ δ(H11O13C6)(10)+ δ(C6H14O13)(10)
1327	1280	0.4	-	ω(C6H8C7)(42)+ ω(N1H4C2)(20)
1309	1262	20.4	1266	ρ(CH₂)(34)+ δ(C15H18O17)(22)+ ρ(CH₂)(15)
1281	1235	24.3	-	ω(CH₂)(29)+ τ(N1H4C2)(15)+ τ(N1H12C3)(13)
1247	1203	0.6	1198	τ(N1H4C2)(24)+ ω(CH₂)(20)+ τ(N1H12C3)(14)+ τ(C6H8C7)(11)
1210	1167	6.1	1178	τ(C6H8C7)(32)+ τ(N1H4C2)(12)
1177	1135	75.4	-	ν(C15O17)(21)+ δ(C15H18O17)(18)+ ν(N1C2)(12)
1173	1131	46.3	-	δ(C6H14O13)(25)+ ω(C2H11C6)(15)+ τ(C2H11C6)(13)
1139	1098	227.5	1089	ν(N1C2)(27)+ ν(N1C2)(14)+ δ(C15H18O17)(12)
1098	1059	29.8	-	ν(C6O13)(20)+ τ(C6H8C7)(14)+ ν(N1C2)(14)+ δinp(R)(13)
1060	1022	2.2	-	τ(N1H4C2)(15)+ ν(C3C7)(11)+ ν(C6O13)(11)
983	948	7.2	-	ν(C3C7)(24)+ ν(N1C2)(13)
977	942	28.7	-	δinp(R)(20)+ ν(C6O13)(13)+ ν(C6C7)(12)+ ν(C3C7)(11)
961	927	14.8	-	δinp(R)(30)+ ν(N1C2)(17)+ ν(C6O13)(14)+ ν(C2C6)(13)
932	899	5.7	-	ν(C6C7)(24)+ ν(C3C15)(16)+ ν(C2C6)(10)
870	839	15.1	-	ν(C2C6)(28)+ ν(C6O13)(17)+ δinp(R)(14)
825	795	30.2	-	δinp(R)(37)+ δoop(C15O16)(18)+ ρ(CH₂)(12)
798	769	26.3	-	δoop(H10C2C3)(29)+ ν(C3C7)(17)
777	749	4.0	-	ν(C3C15)(13)+ δinp(R)(22)+ ν(C6C7)(11)+ ν(C6O13)(10)
722	696	71.8	-	δoop(H10C2C3)(33)+ ν(N1C2)(12)
681	657	84.2	-	δinp(R)(31)+ δoop(C15O16)(20)+ τ(C3C15)(13)+ δoop(H10C2C3)(10)
608	586	33.9	-	δ(O16O17C3)(40)+ τ(C3C15)(28)
549	529	41.5	-	τ(C3C15)(40)+ δoop(C15O16)(15)
454	438	17.1	-	δ(O16O17C3)(29)+ δ(O16C15C3)(16)

Types of vibrations: ν–stretching, ρ–rocking, ω–wagging, τ–twisting, δ–deformation, δinp–in plane ring bending, δoop–out of plane ring bending, τ– ring torsion.

Tables **2** and **3** list the vibrational frequencies (wavenumbers), IR intensity, and corresponding assignments including potential energy distribution (PED) for CHLP and THLP, respectively. The PED analysis is carried out with the *gar2ped* program [18] in this chapter. Note that PED of less than 10% are not included in the assignment. We divide the whole spectra into three regions for the sake of simplicity of discussion.

Spectral Region above 2800 cm⁻¹

In this region, nearly all modes correspond to the stretching vibrations. The highest frequency corresponds to OH stretching with PED 100%, polarized perpendicular to the ring are obtained at 3691cm⁻¹ and 3690 cm⁻¹ in CHLP and THLP, respectively, which is associated to the −OH group, whereas, at 3624 cm⁻¹ in CHLP and 3618 cm⁻¹ in THLP conformer associated to the −COOH group. This difference may be due to the intramolecular hydrogen bond, which is stronger in the case of THLP, 13O-15H (2.23 Å) than *cis* conformer 16O-18H (2.29 Å). At 3421 cm⁻¹, pure NH stretching mode polarized along 3C-15C in *cis* form and corresponding mode in *trans* conformer lies at 3409 cm⁻¹. Furthermore, in CHLP, CH stretching vibrations polarized along a plane contain 7C making 45⁰ with the plane of ring is calculated at 2973 cm⁻¹ with PED of 82%, agreeing well with the experimental value of 2976 cm⁻¹. Similarly, the corresponding mode of vibration polarized along 3C-8H in THLP lies at 2997 cm⁻¹ having an FTIR value of 2986 cm⁻¹. Other CH modes lying at 2966 cm⁻¹ (90% PED) and 2925 cm⁻¹ (84% PED) in CHLP match well with the experimental values, 2976 cm⁻¹ and 2938 cm⁻¹, respectively. Moreover, corresponding modes in THLP, scaled at 2953 cm⁻¹ (65% PED) and 2922 cm⁻¹ (85% PED) are seen at 2951 cm⁻¹ and 2928 cm⁻¹ in FTIR spectrum.

Table 3. Proposed assignments and potential energy distribution (PED) for vibrational modes of THLP calculated at the B3LYP/6-31+G(d,p) level [13].

Calc. Freq. (cm⁻¹)	Scaled Freq. (cm⁻¹)	IR Int. (a.u.)	FTIR Freq. (cm⁻¹)	Assignments (PED ≥10%)
3825	3690	29.0	-	ν(O16H17)(100)
3750	3618	73.0	-	ν(O14H15)(100)
3534	3409	15.7	3286	ν(N1H18)(99)
3112	3009	26.3	-	ν(C7H12)(74) + ν(C7H11)(23)
3107	2997	21.0	2986	ν(C2H4)(65) + ν(C2H5)(33)
3070	2961	12.6	-	ν(C3H8)(91)
3061	2953	40.1	2951	ν(C2H5)(65) + ν(C2H4)(30)
3049	2941	33.3	-	ν(C7H11)(63)+ ν(C7H12)(16)+ ν(C6H10)(11)
3029	2922	29.3	2928	ν(C6H10)(85)+ ν(C7H11)(11)
1811	1747	300.7	-	ν(C9O13)(79)
1505	1452	3.6	-	δ(H4H5C2)(81) + τ(H18C2)(14)
1493	1440	1.6	1430	ν(N1C2)(38)+ δ(H11H12C7)(17)+ δinp(R)(10)
1439	1388	9.0	-	δ(H11H12C7)(56)+ ν(N1C2)(17)

(Table 3) cont.....

Calc. Freq. (cm⁻¹)	Scaled Freq. (cm⁻¹)	IR Int. (a.u.)	FTIR Freq. (cm⁻¹)	Assignments (PED ≥10%)
1409	1359	12.0	1358	δ(H10O61C6)(41)+ ω(NH)(11)
1395	1345	92.6	1339	δ(C12C6O11)(20)+ ν(N1C2)(16)+ ν(C3C9)(10)+ δinp(R)(10)
1357	1309	3.9	1319	ω(NH)(46)
1345	1297	16	-	ω(CH₂)(29)+ ω(N1H8C3)(12)+ τ(C3H8N1)(11)
1324	1277	9.6	-	ω(N1H8C3)(25)+ τ(C3H8N1)(18)+ ω(C6C2H10)(15)
1309	1262	27.4	1258	δinp(R)(18)+ δ(C12C6O11)(17)+ δ(C12C6O11)(11)+ δinp(H18N1)(11)
1262	1217	15.3	1226	τ(C2N1H4)(42)
1260	1215	33.4	1216	ω(CH₂)(22)+ δ(C12C6O11)(20)
1226	1182	43.6	-	δinp(H18N1)(35)+ δoop(N1H18)(15)+ ν(N1C2)(13)
1195	1152	5.0	-	ω(C6C2H10)(20)+ τ(C6C2H10)(15)+ τ(C7C3H11)(14)+ δ(C6H17O16)(11)
1163	1122	211.4	-	ν(C9O14)(32)+ δ(C12C6O11)(24)+ δ(C12C6O11)(10)
1137	1096	100.7	1087	ν(N1C3)(40)+ δ(N1C3C7)(31)+ ν(N1C2)(15)
1085	1046	15.0	-	δinp(H18N1)(17)+ δinp(R)(13)+ ν(C3C7)(11)+ ν(N1C3)(11)
1044	1007	43.2	-	δinp(H18N1)(35)+ δoop(N1H18)(15)+ ν(N1C2)(13)
1039	1002	5.7	-	ν(C6C7)(31)+ δinp(H18N1)(10)
984	949	4.0	-	ν(C3C9)(24)+ δinp(H18N1)(12)+ ν(N1C2)(12)+ ν(C3C7)(10)
964	930	48.1	920	ν(C2C6)(21)+ δinp(R)(19)+ ν(C6O16)(14) + ν(N1C3)(10)+ τ(H18C2)(10)
930	897	10.8	-	ν(C3C7)(16)+ δinp(R)(12)
875	844	2.4	844	δinp(R)(20)+ δinp(H18N1)(16)+ ν(C3C7)(11)+ ν(N1C3)(11)
841	811	5.6	-	ν(C2C6)(22)+ δinp(H18N1)(17)+ ν(C6C7)(12)+ ν(C6O16)(10)
787	759	134.2	-	δoop(N1H18)(59)+ δinp(H18N1)(23)
738	712	31.7	-	ω(C9C3)(34)+ δinp(H18N1)(13)+ δoop(N1H18)(10)
715	689	13.8	-	δoop(N1H18)(26)+ τ(H18C2)(19)
658	634	28.3	-	δ(N1C3C7)(34)+ ρ(O13C3C9)(28)+ δinp(H18N1)(12)
613	591	49.5	-	τ(O14C9)(26)+ δ(N1C3C7)(24)+ δinp(R)(16)+ δinp(H18N1)(10)
593	572	30.0	-	τ(O14C9)(24)+ δinp(H18N1)(19)+ δinp(R)(12)+ δ(N1C3C7)(10)
450	434	25.4	-	δinp(R)(26)+ ρ(O13C3C9)(13)+ δinp(R)(12)
425	410	10.1	-	τ(C6C2H10)(33)+ τ(H18C2)(16)+ ω(C6C2H10)(13)+ δinp(R)(10)

Spectral Region 1800–1000 cm⁻¹

Carbonyl absorptions are very sensitive as both the carbon and oxygen atoms of the −COOH group move during the vibration, having nearly equal amplitude. In CHLP, a very intense band due to CO stretching polarized along the 1N-10H direction is calculated at 1746 cm^{-1} with a PED of 78%. A mixing of vibrational modes obtained at 1322 cm^{-1}, polarized along the plane of the ring, corresponds to the out-of-plane CNH bending (21%) and CH$_2$ rocking (12%), which is in agreement with the experimental one (see Table **2**). At 1291 cm^{-1}, a moderately intense mode corresponding to the mixing of CHC wagging (29%), CC stretching (12%) and CHO out-of-plane bending (10%) matches well with the experimental value of 1298 cm^{-1}. Furthermore, the mixing of −COOH group deformation with CH$_2$ rocking and CN stretching modes is calculated at 1262 cm^{-1} and 1098 cm^{-1} against the experimental bands at 1266 cm^{-1} and 1089 cm^{-1}, respectively.

In THLP, like CHLP, a very intense band due to CO stretching polarized along the 1N-18H direction is calculated at 1747cm^{-1} with a PED of 79%. At 1345 cm^{-1}, a mixing of out-of-plane deformation of −COOH group (20%), CC stretching (16%) and CN stretching (10%) correspond to the experimental band at 1339 cm^{-1}. Other significant modes calculated at 1262 cm^{-1} and 1096 cm^{-1} correspond to the mixing of −COOH group deformation with in-plane NH bending and CCN deformation with CN stretching agree well with the experimental values of 1258 cm^{-1} and 1087 cm^{-1}, respectively.

Spectral Region below 1000 cm^{-1}

As expected, the ring torsions along with wagging modes appear in this low-frequency region. For CHLP, a mixing of modes appears at 696 cm^{-1}, corresponding to out of plane CCH bending (33%) and CN stretching (12%). A polarized mode of vibration with polarization vector along CN bond occurs at 657 cm^{-1}, corresponding to deformations of ring and out-of-plane −COOH vibration with the PED of 31% and 16%, respectively. For THLP, a mixing of modes appears at 689 cm^{-1}, which corresponds to the out-of-plane NH bending (26%) and CH torsion (19%). The deformation of the −COOH group mixed with in-plane NH bending is obtained at 634 cm^{-1}.

Electronic Properties

Total electronic energies (including zero-point energy), energy-related parameters, and dipole moments of CHLP and THLP are collected in Table **4**. THLP conformer is merely 0.027 eV higher in energy than the CHLP. Ionization potential (I), electron affinity (A), and energy gap (E_{gap}) of molecules are calculated as described previously (see Chapter 2).

I and A values of THLP are slightly larger than those for CHLP (see Table **4**). This may suggest that the *trans* conformer is relatively more stable against the addition or removal of an electron. This fact is further supported by its higher E_{gap} as compared to CHLP. More interestingly, the dipole-moment (μ) of THLP exceeds that of CHLP by 1.11 Debye. The higher dipole-moment of *trans* conformer results due to the fact that the groups attached to the ring lie in different planes which increases the polarity, unlike CHLP in which both groups lie in the same plane.

Table 4. Electronic parameters of CHLP and THLP.

Parameters	CHLP	THLP
Total energy (a.u.)	-476.2604	-476.2594
I (eV)	6.35	6.50
A (eV)	0.38	0.47
E_{gap} (eV)	5.97	6.03
μ (Debye)	2.06	3.17

CIS-4-HYDROXY-*D*-PROLINE AND ITS DIMER

Molecular Geometry and Potential Energy Surface

The optimized structure of CHDP is displayed in Fig. (**4**). Like CHLP, it also contains a pentagonal heterocyclic ring functionalized by −OH as well as −COOH groups. The ring is not planar with the dihedral C3-C7-C6-C2 of 31^0 because of the electronic repulsion of the lone-pair of nitrogen. The ring C6−C7 bond length (1.53 Å) is increased only slightly to 1.55 Å (C3-C7) in the vicinity of nitrogen (N1) substituted in the ring.

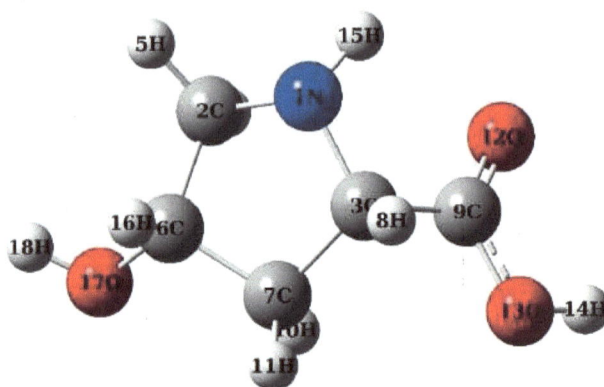

Fig. (4). Global minimum structure of CHDP [14].

To confirm whether the optimized geometry corresponds to the global minimum in the potential energy surface (PES) of the molecule, we have rotated the C3-C9 bond relative to the ring plane, which causes to change the dihedral C6-C3-C9-O12 (which is calculated to be 80^0 in Fig. **4**). Hence, we performed the scanning of the PES with respect to dihedral C6-C3-C9-O12 and displayed the PES scan curve in Fig. (**5**).

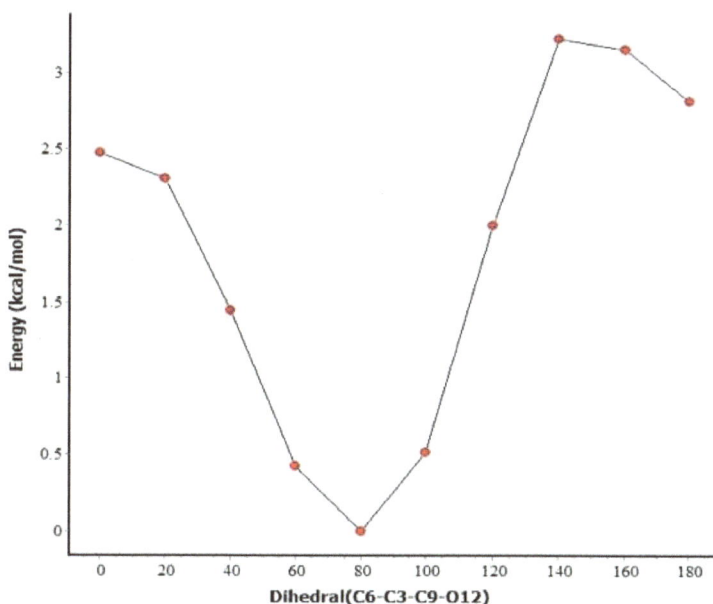

Fig. (5). PES scan curve of CHDP with respect to dihedral (C6-C3-C9-O12) [14].

The PES scan suggests that equilibrium geometry (Fig. **4**) really belongs to the global minimum. The CHDP conformers found are shown in Fig. (**6**) and respective parameters are collected in Table **5**. It can be seen that there are two low-lying conformers, "d" and "e" 'of CHDP. The conformers "d" and "e" possess dihedrals of almost 60^0 and 100^0 as well as relative energies 0.4 and 0.5 kcal/mol, respectively. The dihedrals of the ring of these conformers are approximately same as that of the global minimum structure (having the difference of 1 to 2^0).

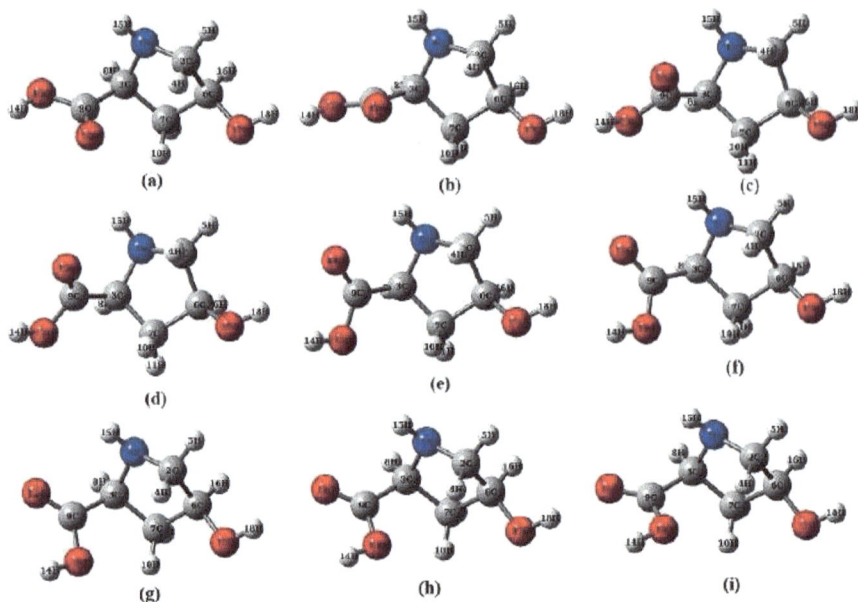

Fig. (6). The CHDP conformers found by PES scan with respect to C6-C3-C9-O12 dihedral [14].

The relative populations of these conformers are obtained by using the Boltzmann distribution as follows,

$$N_f = \frac{e^{-\Delta E/RT}}{\sum_n e^{-\Delta E_n/RT}}$$

(1)

Here ΔE is the energy of a conformer relative to the global minimum structure and summation takes place for all conformers (n). With $R = 1.987 \times 10^{-3}$ kcal/mol·K and $T = 298$ K, we calculated, N_f (d) = 0.24 and N_f (e) = 0.21. Thus, the relative populations of conformers "d" and "e" are 24% and 21%, respectively, but that of the global minimum structure is 47%.

Table 5. Analysis of CHDP conformers at B3LYP/6-31+G(d,p) level [14].

Dihedral	Conformer	ΔE	Ring Dihedral	N_f
(C6-C3-C9-O12)	(Fig. 3)	(kcal/mol)	(C3-C7-C6-C2)	(%)
0	a	2.5	27	0.7
20	b	2.3	28	1.0
40	c	1.5	29	3.9

(*Table 5*) *cont.....*

Dihedral	Conformer	ΔE	Ring Dihedral	N_f
60	d	0.4	30	24
80	-	0	31	47
100	e	0.5	28	21
120	f	2	27	1.7
140	g	3.2	19	0.2
160	h	3.2	21	0.2
180	i	2.8	24	0.4

Frontier Orbitals, MESP Surfaces and Charge Distribution

The HOMO and LUMO surfaces of CHDP are plotted in Fig. (**7**) with corresponding energies calculated relative to the ionization continuum. It is evident that the HOMO is primarily located on the upper half of the ring including nitrogen, but the LUMO is contributed by the side chains of CHDP. Therefore, the transition from HOMO to LUMO corresponds to the transfer of charge from the ring to the side chains. The molecular electrostatic potential (MESP) of CHDP is also displayed in Fig. (**7**), which is mapped surface over uniform electron density. The MESP surface is employed to visualize the molecular size, shape as well as positive and negative electrostatic potential regions at the same time using the color-coding scheme (see Chapter 2). For CHDP, this color-code varies from +0.062 a.u. for deepest blue to -0.062 a.u. for deepest red and (see Fig. **7**). We can note that the hydrogen atoms of the side chains possess the highly electropositive potentials. Therefore, these regions of the side chains are more suitable for the nucleophilic substitution as compared to the ring.

The atomic charge distribution of a molecule is a property that can be analyzed only by theoretical means. The atomic charges of CHDP are obtained using the Mulliken population analysis (MPA) as well as natural population analysis (NPA) methods. MPA is more common and NPA is more accurate because of its low basis set dependency. We have already established the accuracy of NPA over several population schemes in a previous study [19]. The atomic charges of CHDP are plotted in Fig. (**8**). It can be seen that the magnitudes of NPA charges are higher than those of MPA charges because of the fact that the NPA takes the maximum possible occupancy of each atom into account. The charges on carbon atoms are both positive and negative such that their maximal value is for positively charged C9. All O and N atoms carry negative charges and consequently, they are likely to accept electrons. One should also note that hydrogen atoms possess only positive charges. This results in the charge transfer from H to N, C, as well as O atoms.

HOMO Energy = -0.23815 a.u.

LUMO Energy = -0.01736 a.u.

-0.062   0.062

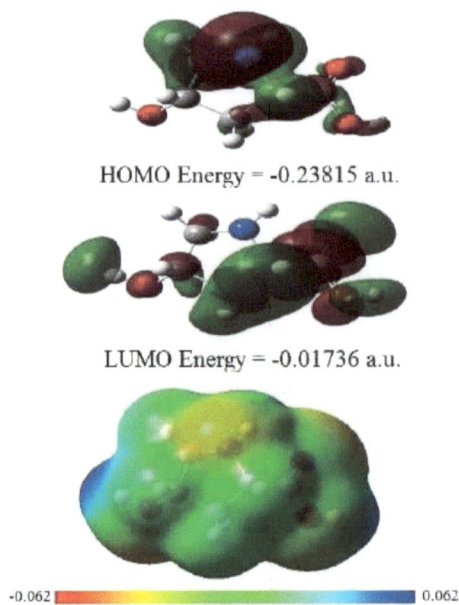

Fig. (7). HOMO, LUMO and MESP surfaces of CHDP [14].

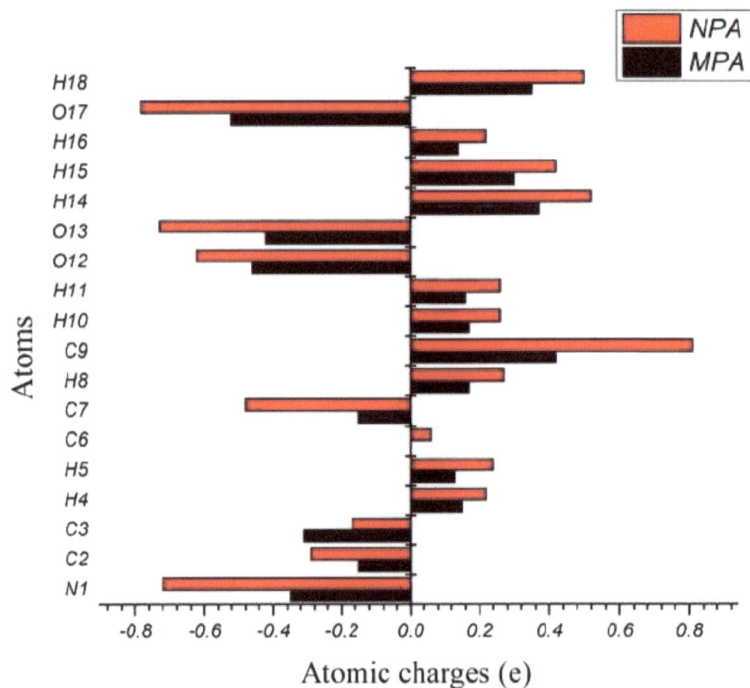

Fig. (8). Atomic charges on CHDP using MPA and NPA schemes [14].

Inter-Molecular Interaction and NBO Analyses

The derivatives of carboxylic acid usually have dimeric character, which can be formed by strong inter-molecular hydrogen bonding interactions in the solid and liquid states. To explore these inter-molecular interactions, we have considered the dimerization of CHDP molecule. The equilibrium geometry of the CHDP dimer is shown in Fig. (**9**). There exist two conformers of CHDP dimer in which both CHDP molecules interact with each other in parallel position (Fig. **9a**) and perpendicular position (Fig. **9b**). The dimer "b" is higher in energy than the conformer "a" by 2.3 kcal/mol. The binding energies of dimeric forms "a" and "b" are calculated to be 14.3 and 12.0 kcal/mol, respectively.

Fig. (9). Equilibrium geometries of CHDP dimer with inter-molecular interactions in dotted green lines [14].

In order to obtain more insights into inter-molecular interactions, we have performed QTAIM (which will be discussed in Chapter 7) and NBO analyses on both dimeric forms. The NBO analysis is widely known to be capable of predicting the hybridization of atoms and lone-pairs contributing in bonding orbital. This analysis can be successfully applied for H-bonded and other van der Waal bonded complexes. The donor (filled) NBO σ of the Lewis structure is well adapted to describe covalency effects while noncovalent delocalization effects are associated with σ → σ* interactions where σ* denotes acceptor (empty) NBO of the non-Lewis structure. The donor-acceptor interactions can be described by the second-order Fock matrix in the NBO basis. Such interactions lead to the charge transfer (CT) from Lewis into an empty non-Lewis orbital. We have exploited NBO analysis to analyze the stabilization in CHDP dimer (a and b) due to interactions of lone pair (n) donor and antibond (σ^*). Table **6** gives the second-order perturbation theory analysis of n→σ^* interactions in the NBO basis. The analysis estimates n→σ^* delocalization by considering energy lowering due to the perturbation and the magnitude of the Fock interaction matrix element (F_{ij}) from donor to an acceptor (see Chapter 2). It is evident that the maximum stabilization (13.56 kcal/mol) in CHDP dimer "b" is due to $n_1(O32)$, $n_2(O32)$ → $\sigma^*(O12\text{-}H13)$ interactions but the similar interaction in "a", $n_2(O11)$ → $\sigma^*(C8\text{-}O12)$ causes the stabilization by 18.8 kcal/mol. This finding is in accordance with the relatively higher energy of inter-molecular interactions in dimer "a" as calculated by the QTAIM (see Chapter 7).

Table 6. The NBO analysis for various lone pairs' interactions in CHDP dimer [14].

a			b		
Donor (*i*)	**Acceptor (*j*)**	**$E^{(2)}$ (kcal/mol)**	**Donor (*i*)**	**Acceptor (*j*)**	**$E^{(2)}$ (kcal/mol)**
$n_1(O34)$	$\sigma^*(C31\text{-}O35)$	6.71	$n_1(O11)$	$\sigma^*(C22\text{-}H28)$	0.71
$n_2(O34)$	$\sigma^*(C31\text{-}C32)$	18	$n_2(O11)$	$\sigma^*(C22\text{-}H28)$	0.73
$n_2(O34)$	$\sigma^*(C31\text{-}O35)$	18.33	$n_1(O11)$	$\sigma^*(C23\text{-}H26)$	0.27
$n_1(O11)$	$\sigma^*(C8\text{-}O12)$	6.54	$n_2(O11)$	$\sigma^*(C23\text{-}H26)$	0.08
$n_2(O11)$	$\sigma^*(C8\text{-}O3)$	17.71	$n_1(O32)$	$\sigma^*(O12\text{-}H13)$	9.51
$n_2(O11)$	$\sigma^*(C8\text{-}O12)$	18.75	$n_2(O32)$	$\sigma^*(O12\text{-}H13)$	13.56

Vibrational Spectroscopic Analysis

As mentioned in Chapter 3, one of the strengths of DFT is the accurate prediction or explanation of the infrared spectra of molecules. Smaller differences between calculated (scaled) and experimental wavenumbers are because of the fact that the FT-IR spectra are determined by dissolving the solid sample in some solvents. There are two general techniques used to obtain the FT-IR spectra- KBr disc, and Nujol mull. In the KBr disc, the peaks at ~3450 cm^{-1} and ~1640 cm^{-1} are very common because of the moisture in the sample. The spectra recorded in the Nujol mull are generally free of such undesired peaks. However, the CH bands due to Nujol (~2950 and 1400 cm^{-1}) must be paid attention during vibrational assignments. In contrast to the FT-IR spectra, the vibrational spectra are calculated for a single molecule (in the gas phase), thus, ignoring all kind of inter-molecular interactions.

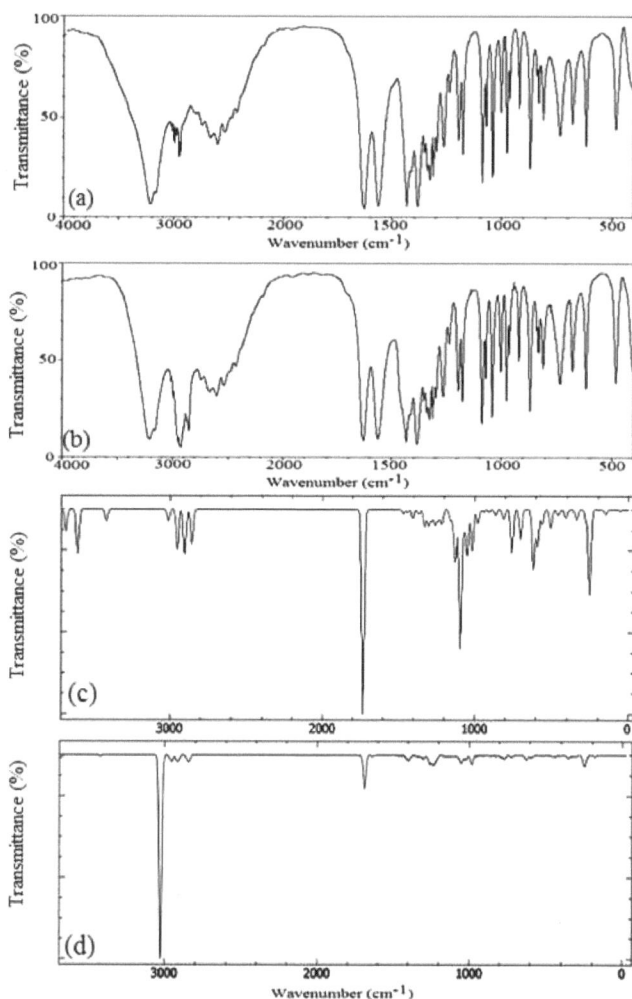

Fig. (10). Experimental FT-IR spectra **(a)** KBr disc, **(b)** Nujol mull, and calculated IR spectra **(c)** monomer, **(d)** dimer of CHDP [14].

We have calculated the vibrational spectra of CHDP as well as its dimer in order to account for intermolecular interactions. The calculated spectra of CHDP monomer and dimer are displayed in Fig. (**10**) along with the FT-IR spectra [17] obtained by KBr disc and Nujol mull techniques for comparison. This enables us to provide more reliable and accurate assignments to the normal modes of CHDP. In Table **7**, we have listed the calculated and scaled wavenumbers with corresponding intensities of CHDP monomer and dimer. All normal modes of vibrations up to 400 cm^{-1} have been assigned based on the PED and compared with respective FT-IR wavenumbers. For clarity of discussion, we classify the vibrational modes into two broad categories;

Ring Vibrations

The ring vibrations usually contain CH stretching, CC stretching, and bending modes. Also, a few modes corresponding to the substituent (N) were also found with significant intensity. CH vibrations are generally observed in the range 3100-3000 cm^{-1}, which are not much affected by a substituent as mentioned in the previous chapter. The IR active bands at 2993-2888 cm^{-1} are assigned to pure CH stretching modes, whereas in dimer these are found at 3134-3055 cm^{-1}. This is in accordance with the corresponding observed values (see Table 7). Most CH stretching modes are weak due to the transfer of charge from hydrogen to carbon atoms, as discussed in an earlier section. Pure NH stretching is obtained at 3393 cm^{-1} in CHDP monomer and 3408-3417 cm^{-1} in CHDP dimer. The modes calculated at 2885 cm^{-1} (monomer) and 2909-2914 cm^{-1} (dimer) are due to the mixing of CH and CN stretchings. These values are in agreement with the FT-IR values of 2938 cm^{-1} (KBr) and 2935 cm^{-1} (Nujol).

The vibrational modes corresponding to CHH, CCH, CNC, *etc.* in-plane and out-of-plane bendings appear below 1500 cm^{-1}. For example, some out-of-plane bending vibrations at 1398 and 1235 cm^{-1} in CHDP monomer are obtained to be at 1456 and 1288 cm^{-1} in its dimer. Moreover, in monomer, in-plane bending vibrations are found at 1358 and 1320 cm^{-1} while in the dimer, these are at 1421 and 1427 cm^{-1}. In the lower wavenumber region, some torsion modes, ring deformation modes, and several mixed vibrations can be obtained. Two strong torsion modes having significant PED contributions in monomer are assigned at 695 and 586 cm^{-1} while that of the dimer at 725 and 618 cm^{-1}.

Table 7. Vibrational analysis of CHDP with frequencies in cm^{-1} and IR intensities in a.u. [14].

Monomer			Dimer						FT-IR Value		Assignments* (PED ≥10%)
Cal.	Scal.	IR Int.	Cal.	Scal.	IR Int.	Cal.	Scal.	IR Int.	KBr Disc	Nujol Mull	
3831	3656	32.9	3831	3644	37.1	3831	3644	28.0	-	-	$v(OH)_{G1}$ (98)
3749	3578	66.2	3152	3002	4249.6	3152	3002	4249.6	-	-	$v(OH)_{G2}$ (99)
3555	3393	17.1	3568	3395	5.86	3559	3386	15.2	-	-	$v(NH)$ (91)
3136	2993	17.6	3147	2997	4.9	3138	2988	65.2	3212	3212	$v(CH)$ (97)
3076	2935	15.5	3090	2943	20.4	3080	2933	35.9	3016	3016	$v(CH)$ (88)
3072	2932	45.3	3073	2927	75.8	3037	2892	32.3	2976	2995	$v(CH)$(98)
3026	2888	19.6	3026	2883	43.2	2965	2825	49.3	2951	2962	$v(CH)$ (96)
1806	1723	300.9	1758	1684	700.9	1703	1632	23.0	1629	1629	$v(CO)_{G2}$ (76)
1529	1459	5.4	1491	1432	1.8	1471	1413	35.3	1564	1563	$v(CO)_{G2}$ (75) + δ(COO) (25)

(Table 7) cont.....

Monomer			Dimer						FT-IR Value		Assignments* (PED ≥10%)
Cal.	Scal.	IR Int.	Cal.	Scal.	IR Int.	Cal.	Scal.	IR Int.	KBr Disc	Nujol Mull	
1465	1398	3.6	1456	1398	94.5	1442	1385	14.9	1433	1434	ω (CHH) (68)+ δ(CHH) (23)
1423	1358	12.4	1422	1366	3.6	1420	1364	4.8	-	-	δ(CHH)(65) + ω(CHH) (25)
1383	1320	6.1	1392	1338	27.8	1371	1318	23.6	1384	1384	δ(CCH) (87)
1357	1295	24.9	1355	1303	50.7	1355	1303	13.2	1353	1354	δ(CHH) (66)
1336	1275	23.2	1348	1296	2.7	1348	1296	21.7	1313	1313	δ(NCC) (54) + ω(CHC) (10)
1313	1253	14.4	1313	1263	109.2	1308	1259	59.2	1298	1298	δ(NCH) (27) + δ(CCH)(25)
1294	1235	21.1	1292	1243	112.5	1286	1238	39.3	-	-	δ(OCH)$_{G2}$ (17)
1285	1227	11.8	1282	1234	32.0	1275	1227	80.5	-	-	δ(CNC)(40) + ν(CC) (33)
1266	1208	5.0	1263	1216	23.9	1262	1215	28.2	1266	1266	δ(CNC)(46) + ν(CC) (34)
1207	1152	21.2	1211	1167	8.8	1207	1163	16.3	1198	-	δ(CNC)(27) + ν(CC) (25) + ν(CO)$_{G2}$ (18)+ δ(OCH)$_{G2}$ (10)
1184	1129	15.6	1185	1142	8.2	1182	1139	21.9	1178	1178	δ(CNC)(15) + δ(CHO)$_{G2}$ (12) + ν(CC) (12)
1172	1119	20.4	1156	1114	25.2	1133	1093	18.0	-	-	δ(CNC)(18) + δ(CCH) (13) + ν(NC) (13) + δ(CCC) (10)
1117	1066	202.5	1116	1077	32.5	1101	1063	62.8	1089	1089	ν(NC)(25)
1094	1044	27.6	1096	1058	48.6	1095	1057	60.7	1071	1071	δ(R2)(19)+ ν(CC) (15)
1060	1011	66.6	1072	1035	48.2	1064	1028	58.2	1041	1042	δ(CNC)(16) + ω(CO)$_{G2}$ (12)
1022	975	60.9	1022	988	25.4	1020	986	22.5	1003	1003	δ(CNC)(22) + ν(NC) (6)
981	936	20.1	993	961	1.3	986	954	10.7	976	976	ν(NC) (21) + ν(CC) (25)
944	901	2.8	949	919	5.1	945	915	4.9	920	-	ω(NHH) (10)
901	860	3.4	901	874	8.0	891	864	8.8	869	870	τ(R2)(11)
845	806	13.7	860	835	11.7	849	825	7.2	811	811	τ(R2)(59)
788	752	63.6	822	799	35.2	802	780	72.2	736	734	ω(COH)$_{G2}$ (36) + δ(R1) (28)
729	695	44.3	755	736	34.4	742	724	7.2	681	681	τ(R2)(22)+ δ(COO)$_{G2}$ (11)

(Table 7) cont.....

Monomer			Dimer						FT-IR Value		Assignments* (PED ≥10%)
Cal.	Scal.	IR Int.	Cal.	Scal.	IR Int.	Cal.	Scal.	IR Int.	KBr Disc	Nujol Mull	
641	611	87.7	653	639	88.0	646	633	8.7	620	620	δ(CCC)(17) + δ(R1)(15) + ω(COH)$_{G2}$ (12)
614	586	53.1	621	609	50.1	602	591	14.4	-	-	τ(R2)(45) + τ(R1)(16)
584	557	21.6	558	550	8.3	535	528	10.1	-	-	τ(R2)(25)
526	502	27.0	495	490	7.4	479	475	10.0	483	482	ω(COH)$_{G2}$ (41) + ρ(OCO) (10)
476	454	8.3	460	457	25.7	427	425	11.4	-	-	ω(CHC)(19) + δ (OCH)$_{G2}$ (17)

Abbreviations: ν–stretching, ρ–rocking, ω–wagging, δ–bending, τ–torsion, R1, R2 – rings, G1, G2 –OH and COOH groups.

Groups Vibrations

CHDP possesses two groups −OH (G1) and −COOH (G2) substituted at the ring. The OH stretching of G1 is found at 3656 cm^{-1} in the monomer, but within 3677-3688 cm^{-1} in the dimer. Further, the **OH** stretching vibration of the hydroxyl group is dependent on the hydrogen bonding. In the gas phase, alcohols give a sharp peak in the region between 3620 and 3670 cm^{-1}, whereas hydrogen-bonded OH stretching appears in the range 3500-3200 cm^{-1}. Thus, the calculated wavenumbers of CHDP indicate that OH of G1 is not involved in hydrogen bonding, which is evident from its dimeric structure (see Fig. **9**).

The vibrations of G2 incorporate CO and OH stretchings along with COH and COO bending. These OH bands are obtained to be at lower wavenumbers and broader than alcohols because of their participation in hydrogen bonding. For example, OH stretching of G2 is obtained at 3578 cm^{-1} (monomer) and 3597-3580 cm^{-1} (dimer). The CO absorption peaks appear in the region 1450–1800 cm^{-1}. Two strong stretching modes of CO are found at 1723 and 1459 cm^{-1} in monomer and 1721 and 1479 cm^{-1} in CHDP dimer. Other vibrations corresponding to in-plane and out-of-plane bending of G2 are found in the middle and lower wavenumber region.

Electronic and Thermodynamic Parameters (Monomer and Dimer)

The HOMO and LUMO energy eigenvalues of molecules along with their energy gap (E_{gap}) help to determine their chemical reactivity as well as kinetic stability. The larger E_{gap} corresponds to the higher kinetic stability due to the fact that it is energetically favorable neither to attach electrons to a high-lying LUMO nor to detach electrons from a low-lying HOMO in order to form the activated complexes in any chemical reaction [20]. The dipole moment (μ) of a molecule provides a stamp of its geometry as well as charge distribution such that the molecules with larger μ are considered to be more polarizable.

Table 8. Electronic and thermodynamic parameters of CHDP monomer and dimer [14].

Electronic Parameter	Monomer	Dimer	Thermodynamic Parameter	Monomer	Dimer
I (eV)	6.48	6.47	ZPE (kcal/mo	93.3	187.7
A (eV)	0.47	0.73	E (kcal/mol)	98.8	199.5
E_{gap} (eV)	6.01	5.75	C_v (cal/mol-K)	32.3	68.1
μ (Debye)	1.11	1.92	S (cal/mol-K)	93.4	149.8

Table **8** lists the calculated electronic parameters of CHDP monomer as well as dimer. Note that ionization potential (I), electron affinity (A), and E_{gap} of CHDP are slightly higher than those of CHLP (see Table **4**). Further, I and A of CHDP dimer are a bit lower and larger than those of its monomer, respectively. This should suggest more chemically reactive nature of the CHDP dimer. This result is in agreement with its smaller E_{gap} and larger μ values. Various thermodynamic parameters namely the zero-point energy (ZPE), thermal energy (E), constant volume heat capacity (C_v), and entropy (S) for CHDP monomer and dimer are also given in Table **8**. These parameters are mutually related by standard thermodynamic equations, which may be important in the study of paths of chemical reaction.

CONCLUDING REMARKS

Using B3LYP/6-31+G(d,p) calculations, we have discussed the structural, vibrational, and electronic properties of 4-hydroxy-*l*-proline in *cis* and *trans* conformations. Detailed assignments to all normal modes up to 450 cm^{-1} have been presented and potential energy distribution along with the direction of polarization has also been discussed. The calculated vibrational wavenumbers were scaled by a scale factor and found in better agreement with corresponding FT-IR spectra. Furthermore, we have noticed that the *trans* conformer is relatively more stable and polar as compared to its *cis* counterpart.

A descriptive study on *cis*-4-hydroxy-*d*-proline (CHDP) has also been discussed. The global minimum of CHDP structure has been achieved by scanning the potential energy surface. The frontier orbitals and MESP surfaces are employed to discuss the chemical reactivity and partial atomic charges are computed by using the Mulliken and natural population schemes. The optimized structure of CHDP dimer has been obatined and intermolecular interactions are analyzed using the NBO method. The infrared spectral analysis has been carried out on CHDP monomer and dimer to the greater accuracy. The scaled wavenumbers, by a scaling equation, are compared with experimental FT-IR peaks observed by KBr disc and Nujol mull techniques. Several electronic and thermodynamic parameters for CHDP monomer and dimer are also reported.

CONSENT FOR PUBLICATION

Not applicable.

CONFLICT OF INTEREST

The authors declare no conflict of interest, financial or otherwise.

ACKNOWLEDGEMENTS

Declared none.

REFERENCES

[1]   Alberts, B.; Johnson, A.; Lewis, L. *Molecular Biology of the Cell,* 4th ed; Garland Science: New York, **2002**.

[2]   Hohsaka, T.; Sisido, M. Incorporation of non-natural amino acids into proteins. *Curr. Opin. Chem. Biol.,* **2002**, *6*(6), 809-815.
        [http://dx.doi.org/10.1016/S1367-5931(02)00376-9] [PMID: 12470735]

[3]   Dougherty, D.A. Unnatural amino acids as probes of protein structure and function. *Curr. Opin. Chem. Biol.,* **2000**, *4*(6), 645-652.
        [http://dx.doi.org/10.1016/S1367-5931(00)00148-4] [PMID: 11102869]

[4]   Lehninger, A.L.; Nelson, D.L.; Cox, M.M. *Principles of Biochemistry,* 3rd ed; W. H. Freeman: New York, **2000**.

[5]   Pavlov, M.Y.; Watts, R.E.; Tan, Z.; Cornish, V.W.; Ehrenberg, M.; Forster, A.C. Slow peptide bond formation by proline and other N-alkylamino acids in translation. *Proc. Natl. Acad. Sci. USA,* **2009**, *106*(1), 50-54.
        [http://dx.doi.org/10.1073/pnas.0809211106] [PMID: 19104062]

[6]   Gorres, K.L.; Raines, R.T. Prolyl 4-hydroxylase. *Crit. Rev. Biochem. Mol. Biol.,* **2010**, *45*(2), 106-124.
        [http://dx.doi.org/10.3109/10409231003627991] [PMID: 20199358]

[7]   Szpak, P. Fish bone chemistry and ultrastructure: implications for taphonomy and stable isotope analysis. *J. Archaeol. Sci.,* **2011**, *38*, 3358-3372.
        [http://dx.doi.org/10.1016/j.jas.2011.07.022]

[8]   Nelson, D.L.; Cox, M.M. *Lehninger's Principles of Biochemistry,* 4th ed; W. H. Freeman: New York, **2005**.

[9]   Brinckmann, J.; Notbohm, H.; Müller, P.K. *Collagen: Topics in Current Chemistry 247*; Springer: Berlin, **2005**.
        [http://dx.doi.org/10.1007/b98359]

[10]   Cassab, G.I. Plant cell wall proteins. *Annu. Rev. Plant Physiol. Plant Mol. Biol.,* **1998**, *49*, 281-309.
        [http://dx.doi.org/10.1146/annurev.arplant.49.1.281] [PMID: 15012236]

[11]   http://www.wheelessonline.com/ortho/pagets_disease

[12]   Wieland, T. *Peptides of Poisonous Amanita Mushrooms*; Springer, **1986**.
        [http://dx.doi.org/10.1007/978-3-642-71295-1]

[13]   Srivastava, A.K.; Pandey, A.K.; Gangwar, S.K.; Misra, N. Structural, vibrational and electronic
        properties of cis and trans conformers of 4-hydroxy-l-proline: a density functional approach. *J. At.
        Mol. Sci.,* **2014**, *5*, 279-288.

[14]   Srivastava, A.K.; Dwivedi, A.; Kumar, A.; Gangwar, S.K.; Misra, N. Conformational analysis,
        intermolecular interactions, electronic properties and vibrational spectroscopic studies on cis--
        -hydroxy-dproline. *Cogent Chem.,* **2016**, *2*, 1149927.
        [http://dx.doi.org/10.1080/23312009.2016.1149927]

[15]   Koetzle, T.F.; Lehmann, M.S.; Hamilton, W.C. Precision neutron diffraction structure determination
        of protein and nucleic acid components. IX. The crystal and molecular structure of 4-hydroxy-
        L-proline. *Acta Crystallogr. B,* **1973**, *29*, 231-236.
        [http://dx.doi.org/10.1107/S0567740873002256]

[16]   Alecu, I.M.; Zheng, J.; Zhao, Y.; Truhlar, D.G. Computational thermochemistry: scale factor databases
        and scale factors for vibrational frequencies obtained from electronic model chemistries. *J. Chem.
        Theory Comput.,* **2010**, *6*(9), 2872-2887.
        [http://dx.doi.org/10.1021/ct100326h] [PMID: 26616087]

[17]   http://sdbs.db.aist.go.jp

[18]   Martin, J.M.L.; Alsenoy, V.; Alsenoy, C.V. *Gar2ped*; University of Antwerp, **1995**.

[19]   Srivastava, A.K.; Misra, N. Structures, stabilities, electronic and magnetic properties of small Rh_xMn_y
        (x+ y= 2–4) clusters. *Comput. Theor. Chem.,* **2014**, *1047*, 1-5.
        [http://dx.doi.org/10.1016/j.comptc.2014.08.008]

[20]   Manolopoulos, D.E.; May, J.C.; Down, S.E. Theoretical studies of the fullerenes: C_{34} to C_{70}. *Chem.
        Phys. Lett.,* **1991**, *181*, 105.
        [http://dx.doi.org/10.1016/0009-2614(91)90340-F]

CHAPTER 5

DFT Study on Some Natural Products: Triclisine, Rufescine, and Imerubrine

Abstract: This chapter deals with three biologically active natural products, triclisine, rufescine, and imerubrine. The B3PW91/6-311+G(d,p) level of DFT is used to obtain the optimized structures of molecules under study. We carried out vibrational analyses of triclisine and rufescine at their optimized structures and provided detailed assignments of the prominent vibrational modes. The computed infrared frequencies are found to be in good agreement with the experimentally determined FT-IR spectra of both molecules. Similarly, their properties are also studied with the help of HOMO, LUMO, MESP surfaces, and several electronic as well as thermodynamic parameters. The vibrational spectrum of imerubrine is analyzed and the normal modes are accurately assigned based on the potential energy distribution. The nuclear magnetic resonance shifts of imerubrine are also obtained, analyzed, and compared with the corresponding experimental values. The chemical reactivity of imerubrine is also explained using HOMO, LUMO, and MESP surfaces as well as a number of reactivity descriptors.

Keywords: Azafluoranthenes, B3PW91, DFT, Electronic parameter, FT-IR, H-bond, HOMO, Imerubrine, LUMO, MESP, Natural product, NMR, Rufescine, Thermodynamic parameter, Triclisine, Vibrational spectra.

INTRODUCTION

Amazonian wines *Abuta rufescens* and *Triclisia gilletii* are employed to extract triclisine and rufescine, respectively, which are natural products belonging to the azafluoranthene group. The azafluoranthenes are present everywhere in nature, having their parent bases in the cigarette smoke, coal tar, street dust, rivers, and lake sediments [1, 2]. These compounds containing nitrogen-heterocycles have become one of the major contaminants in environment [3, 4]. Nevertheless, the natural products of the azafluoranthene group have been found to show potential-biological activities as antifungal, anti-HIV and cytotoxic agents [5 - 7]. For instance, the azafluoranthene eupolauridine exhibit the cytotoxic activity by targeting the DNA topoisomerase II [8]. Apart from the aforementioned biological activities, the molecules with the azafluoranthene core possess interesting spectral properties due to extensive conjugation. This makes them ideal for luminescent

Ambrish Kumar Srivastava and Neeraj Misra

applications and electroluminescent devices [9 - 12]. The direct arylation [13] and electrocyclization [14] can be used to synthesize the azafluoranthene compounds and their derivatives. The comparative DFT study of triclisine and rufescine has been reported by us [15].

Likewise, *Abuta imene* is used to extract imerubrine, which belongs to the rare natural products from tropoloisoquinolines [16]. Biosynthetically, it is an analogue of alkaloids of azafluoranthene [17]. Boger and Takahashi [18] along with Lee and Cha [19] have reported the total syntheses of imerubrine and associated compounds using the cycloaddition reaction. The structure of imerubrine is derived from rufescine such that extra oxygen is attached to a phenyl ring, as studied by us [20]. Hence, it becomes obvious to perform a comparative analysis of imerubrine and rufescine. In this chapter, we present the results of triclisine, rufescine, and imerubrine at the B3PW91/6-311+G(d,p) level.

TRICLISINE AND RUFESCINE

Molecular Structures

The optimized structures of triclisine and rufescine molecules are shown in Fig. (1) with the labeling of atoms. The structural similarities in these compounds is due to the presence of four six-membered rings including a heterocyclic ring (R1) with the substitution of nitrogen and the functionalized ring (R4) having two –OCH$_3$ groups. In addition, there exist two more –OCH$_3$ groups, in rufescine, linked at the C2 and C16 sites of ring R4 and R3, respectively.

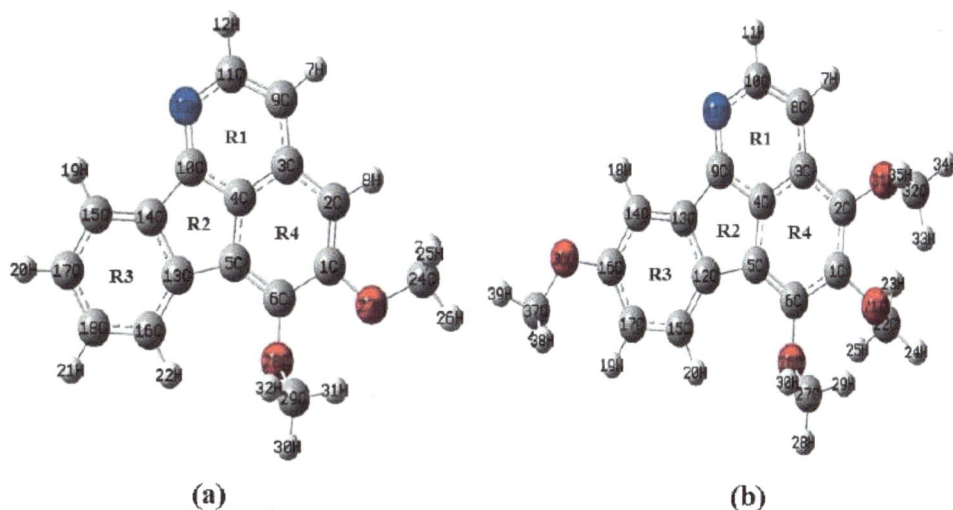

(a) **(b)**

Fig. (1). Optimized molecular geometries of **(a)** triclisine and **(b)** rufescine (reproduced from [15] with the permission of Elsevier).

Infrared Spectroscopic Analysis

Infrared frequencies (wavenumbers) of both molecules are also calculated using the B3PW91/6-311+G(d,p) level. The calculated wavenumbers are scaled with the factor of 0.9648 as discussed in the previous chapters. The Perkin Elmer 1800 Spectrophotometer has been employed for recording the FT-IR spectra of these compounds in the region 4000-400 cm^{-1} using samples in CsI pellet as mentioned earlier [15]. The calculated spectra have been plotted in Fig. (2) along with corresponding FT-IR spectra for a quick comparison.

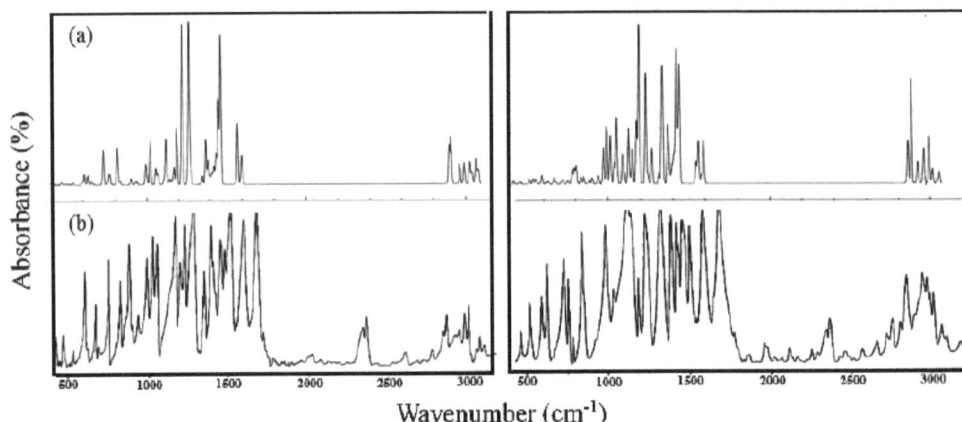

Fig. (2). (a) Simulated IR and (b) experimental FT-IR spectra of (left) triclisine and (left) rufescine (reproduced from [15] with the permission of Elsevier).

Tables 1 and 2 list the prominent vibrational modes for triclisine and rufescine, respectively. The detailed vibrational assignments of normal modes have been offered and respective FT-IR wavenumbers are also listed. To simplify the discussion, the vibrational spectrum has been divided into two parts, greater than 1500 cm^{-1} and less than 1500 cm^{-1}.

Table 1. Infrared spectral analysis of significant vibrations of triclisine (adopted from [15] with the permission of Elsevier).

Wavenumber (cm^{-1})		FT-IR Value (cm^{-1})	IR Int. (a.u.)	Vibrational Assignments*
Calculated	Scaled			
3209	3094	3090	9.3	v(CH)R3
3209	3093	-	6.6	v(CH)R4
3201	3086	-	13.0	v(CH)R3
3191	3076	3080	23.0	v(CH)R1

(Table 1) cont.....

3189	3074	-	12.3	v(CH)R3
3175	3061	-	1.5	v(CH)R3
3165	3051	-	16.3	v(CH)R3
3152	3038	-	19.0	v(24C-26H)
3150	3037	-	12.1	v_{as}(CH$_3$)adj28O
3114	3002	-	29.1	v_{as}(CH$_3$)adj28O
3086	2975	2970	27.8	v_{as}(CH$_3$)adj23O
3030	2921	-	64.2	v(CH$_3$)adj28O
3017	2909	2900	53.1	v(CH$_3$)adj23O
1670	1610	-	8.7	v(CCC)R3R2R4 + β(CH)R4
1665	1605	-	33.8	v(CCC)R1+ β(CH)R1
1659	1599	-	53.1	v(CCC)R1+ v(CN)R1
1639	1580	-	2.9	v(CCC)R3R2+ β(CH)R3
1528	1473	1480	204.0	v(CCC)R1+ β(CH)R4+σ(CH$_3$)adj28O
1513	1458	-	115.4	v(CCC)R1R2+ β(CH)R1 +σ(CH$_3$)adj23O
1503	1448	-	8.4	σ(CH$_3$)adjR4
1498	1444	-	36.5	σCH$_3$)adj28O
1487	1433	-	11.4	σ(CH$_3$)adj23O
1482	1429	-	15.4	σ(CH$_3$)adj23O
1472	1419	-	13.0	ω(CH$_3$)adj28O + β(CH)R3
1445	1393	-	32.8	ω(CH$_3$)adj23O + β(CH)R1R3 + v(5C-4C)
1430	1379	-	61.3	v(1C-6C)R1R2+ β(CH)R3 + v(CN)R1
1408	1357	1352	11.6	β(CH)R1+ β(CCC)R1R4
1329	1281	-	27.0	v(CCC)R1R2R3R4 + β(CH)R1+ v(CN)R1
1318	1271	-	218.8	β(CH)R3R1
1311	1264	-	19.5	β(CH)R3
1297	1250	1252	218.5	β(CH)R1+ω(CH$_3$)adj23O + β(CCC)R1+ β(2C-8H)
1216	1173	-	23.2	ρ(CH$_3$)adjR4+ β(CH)R3R1
1111	1071	-	15.4	β(CH)R3
1089	1050	-	22.4	β(9C-7H) + β(2C-8H)+ v(O-CH$_3$)adj28O
1062	1024	1032	60.1	β(CCC)R3+ β(CH)R1R3+ β(CNC)R1
1033	996	-	27.4	β(CH)R3+ v(O-CH$_3$)adj23O + β(CCC)R3
849	819	815	49.3	γ(CH)R3R4

*Abbreviations: v –stretching; v_{as} – asymmetric stretching; β – in-plane-bending; γ – out-of-plane bending; ω – wagging; ρ – rocking; τ – torsion, σ – scissoring.

Wavenumber Range Above 1500 cm⁻¹

Wavenumber Range Above 1500 cm^{-1}

This range contains various CC and CH stretching modes corresponding to the rings as well as the –OCH$_3$ group. Besides, some in-plane CCC and CCH bending modes have been found. The CH stretching of the rings are normally obtained within the range of 3100-3000 cm^{-1} and these modes are almost unaffected by the nature of substitution. These vibrations lie in the range 3094-3051 cm^{-1} for triclisine and in the range 3111-3065 cm^{-1} in rufescine. These modes show agreement with the determined FT-IR peaks in triclisine (3090, 3080 cm^{-1}) and rufescine (3060 cm^{-1}). Likewise, a CH mode having medium intensity is calculated at 3061 cm^{-1} in triclisine, which is analogous to the mode at 3065 cm^{-1} in rufescine. The largest wavenumber has been noticed for the stretching of CH in ring R3 in both molecules such that the frequencies of rufescine are a bit higher as compared to triclisine, which can be expected due to an electron-withdrawing nature of –OCH$_3$ linked to R3. Similarly, an anti-symmetric stretching of CH$_3$, adjacent to 28O group is obtained at 3037 cm^{-1} in triclisine and the analogous mode of CH$_3$ (adj26O) in rufescine appears at 3010 cm^{-1}. These bands corresponding to the CH$_3$ stretching of the –OCH$_3$ group have been determined at 2970, 2900 cm^{-1} (triclisine) and 3010 cm^{-1} (rufescine) in FT-IR spectra. Similarly, a strong vibrational mode corresponding to CCC stretching in rufescine appears at 1621 cm^{-1}, which is in accordance with the experimental band. The corresponding mode in triclisine is obtained between 1610 and 1605 cm^{-1}, which is slightly lower as compared to rufescine. This can be explained on the basis of the electron deficiencies in the ring R3 in rufescine due to the electron-withdrawing nature of the group linked to the ring.

Wavenumber Range Below 1500 cm⁻¹

Wavenumber Range Below 1500 cm^{-1}

This range covers the bending vibrations corresponding to the rings and –OCH$_3$ groups. This also includes the coupling of several vibrations such as torsion, twisting, wagging, *etc.* To be specific, a strong peak at 1469 cm^{-1} in rufescine results due to the coupling of three vibrational modes, namely, CCC stretching, CH$_3$ rocking, and in-plane CH bending. The corresponding vibrational peak in triclisine has been calculated at 1473 cm^{-1} with the almost intensity. This mode also incorporates the mixing of several vibrational modes including CCC stretching, CH$_3$ scissoring as well as in-plane CH bending, and corresponds to the FT-IR band at 1480 cm^{-1}.

Table 2. Infrared spectral analysis of significant vibrations of rufescine (adopted from [15] with the permission of Elsevier).

Wavenumber (cm⁻¹)		FT-IR Value (cm⁻¹)	IR Int. (a.u.)	Vibrational Assignments
Calculated	Scaled			
3217	3111	-	9.9	ν(CH)R3
3214	3108	-	9.9	ν(CH)R1
3170	3065	3060	27.3	ν(CH)R1
3149	3045	-	11.6	ν_{as}(CH₃)adj31O &26O
3148	3044	-	24.6	ν(CH₃)adj26O
3147	3043	-	26.1	ν(CH₃)adj36O
3144	3040	-	28.9	ν(CH₃)adj21O
3113	3010	3010	53.8	ν(CH₃)adj26O&31O
3011	2911	-	39.4	ν_{as}(CH₃)adj36O
3030	2930	-	59.7	ν(CH₃)adj26O&21O
3029	2929	-	65.5	ν(CH₃) adj21O
3028	2928	-	63.0	ν(CH₃)adj31O
3018	2919	-	76.2	ν(CH₃)adj36O
1677	1627	1624	76.8	ν(CCC)R1R2
1642	1588	-	76.3	ν(CN)R1+ ν(CCC)R1R2+ β(CH)R1
1628	1574	-	39.9	β(CH)R1R3+ β(CCC)R3
1519	1469	-	202.3	ρ(CH₃)adj31O+ β(CCC)R3+ ν(CCC)R3R4
1509	1459	1460	52.2	σ(CH₃)adj31O
1499	1450	-	71.2	σ(CH₃)adj21O
1497	1448	-	131.6	σ(CH₃)adj 36O
1495	1446	-	36.0	σ(CH₃)adj21O
1453	1405	-	44.5	ρ(CH₃)adj36O+ β(CH)R3+ ν(CCC)R1R2&R3
1444	1397	-	98.1	ρ(CH₃)adj31O&21O+ ν(CCC)R4
1412	1365	1370	171.1	ν(CCC)R3+ρ(CH₃)adj31O+ β(CH)R1R3
1405	1359	-	94.8	ν(CCC)R3+ρ(CH₃)adj31O+ β(CH)R1R3
1344	1299	-	62.2	β(CH)R3+ ν(CCC)R4
1303	1260	1258	197.6	ν(1C-6C)+ β(CH)R1
1259	1218	1216	283.2	β(CCC)R1R2R3&R4+ β(CH)R3R1
1245	1204	-	31.9	β(CH)R3R1
1243	1202	-	82.3	β(CH)R3R3+ β(CCC)R4

(Table 2) cont.....

1218	1178	-	60.8	$\rho(CH_3)adj31O\&26O$
1192	1153	-	97.1	$\beta(CH)R3R1+\rho(CH_3)adj31O$
1116	1079	-	106.5	$\beta(CH)R3$
1100	1064	-	38.8	$\beta(8C-7H)$
1075	1039	-	85.9	$\beta(14C-18H)+ \nu(O-CH_3)adj\ 21O$
1052	1017	1014	102.5	$\beta(CNC)R1+ \beta(CH)R1R3$
1030	996	990	63.1	$\beta(CCC)R3+ \beta(CH)R3$
282	273	-	11.1	$\gamma(CH)R1$

The stretching of CN associated with the ring R1 in triclisine is found at 1588 cm^{-1}, which is lower than 1599 cm^{-1} for rufescine. Similarly, the scissoring of CH$_3$ for triclisine (adjacent to 23O) calculated at 1433 cm^{-1} is also lower than the corresponding mode adjacent to 21O in rufescine at 1446 cm^{-1}. Nevertheless, a very strong mode in rufescine at 1260 cm^{-1} corresponds to the CCC stretching associated with the ring R4, which is in excellent agreement with the FT-IR band determined at 1258 cm^{-1}. The vibrational mode at 1250 cm^{-1} in triclisine, however, results due to a mixing of various modes polarized along 1C-5C, which is again in accordance with the corresponding FT-IR band at 1252 cm^{-1}. For even lower wavenumber region, *i.e.*, below 800 cm^{-1}, the coupling of several vibrational modes such as out-of-plane bending, torsions, twisting, *etc.* are found. It is also apparent from the above discussion that there is a good matching between experimental and calculated vibrational wavenumbers in general. One can see that the level of computation employed herein is fully capable of interpreting the observed FT-IR spectra and therefore, it can be a method of choice for spectroscopic analysis of related molecules.

HOMO, LUMO AND MESP SURFACES

The highest occupied molecular orbital (HOMO) and lowest unoccupied molecular orbital (LUMO) of molecular systems are very crucial for the chemical reactivity of molecules as mentioned earlier. These frontier orbitals take part in chemical interactions or their reactions with other systems and the frontier orbitals energy gap (E_{gap}) of a molecule can be used to determine its chemical reactivity [21]. Fig. (**3**) plots the frontier orbitals surfaces of triclisine and rufescine. It can be observed that the ring systems contribute to the HOMOs in both molecules. Likewise, the LUMOs are delocalized over the whole system, except nitrogen atom in rufescine. The E_{gap} of rufescine (3.56 eV) is smaller than that of triclisine (4.01 eV). The smaller E_{gap} implies that the chemical reactivity of rufescine should be enhanced when compared with triclisine. This result is in accordance with the enhanced hydrogen bonding in rufescine as revealed in chapter 7.

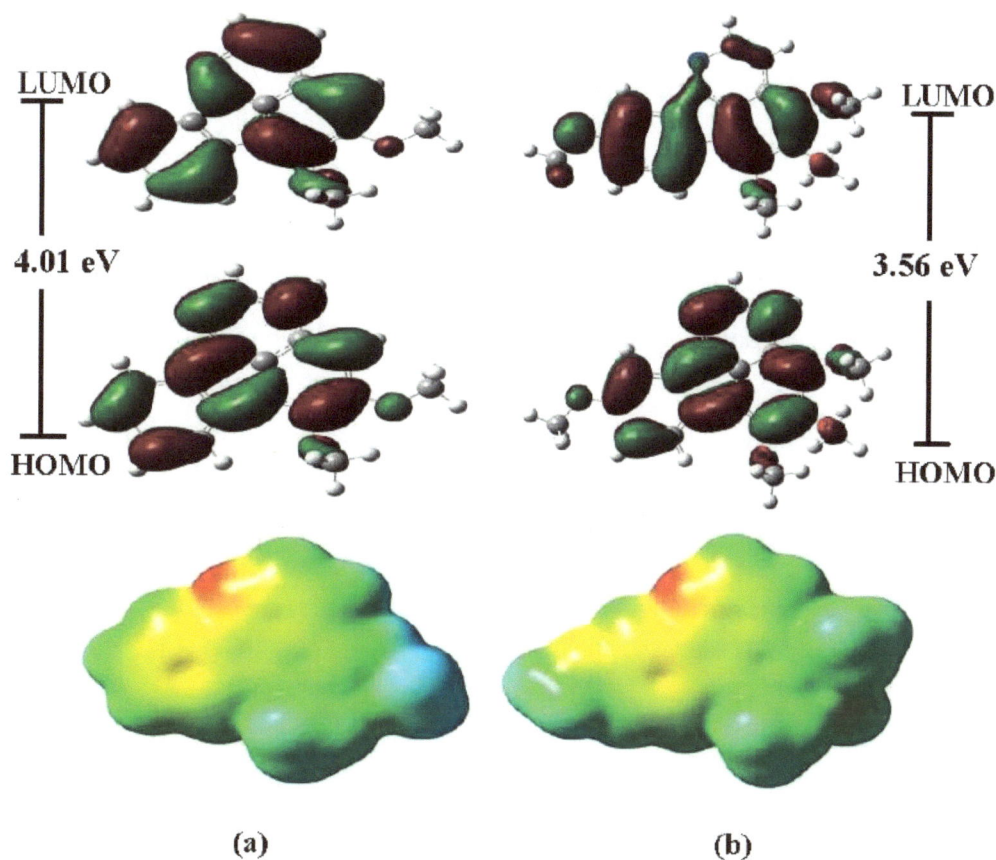

Fig. (3). The frontier orbitals and MESP surfaces of **(a)** triclisine and **(b)** rufescine (reproduced from [15] with the permission of Elsevier).

The map of electrostatic potential on the total density surface of molecular systems gives the molecular electrostatic potential (MESP). The MESP is used to display the variation of electrostatic potential from positive to negative regions simultaneously with the molecular size and shape. Its importance lies in the elucidation of the most favorable sites for molecules in the binding to a receptor along with their geometry (position) [22, 23]. The MESP surfaces of both molecules are also plotted in Fig. (**3**), in which the blue corresponds to the highly positive, *i.e.*, the most electron-deficit region and the red represents the highly negative region, *i.e.*, the most electron-rich region. One can note that the vicinity of the nitrogen atom in the ring R1 is highly electronegative, which can promptly act like a donor of electron in both triclisine and rufescine as expected.

Electronic and Thermodynamic Parameters

The electronic properties of both molecules have been discussed by calculating several electronic parameters, namely, ionization potential (I), electron affinity (A), absolute electronegativity (χ) and chemical hardness (η), *etc.* (see Chapter 2) at B3PW91/6-311+G(d,p) level. These parameters are generally employed to discuss the chemical reactivity of molecular systems, as listed in Table **3**. It is evident from Table **3** that I and A values of rufescine are a bit lower as compared to triclisine, and consequently, its electronegativity. It can also be added that rufescine is chemically softer than triclisine because of enhanced hydrogen bonding (see Chapter 7).

Table 3. The electronic and thermodynamic parameters of triclisine and rufescine (reproduced from [15] with the permission of Elsevier).

Parameters	Triclisine	Rufescine	Parameters	Triclisine	Rufescine
I (eV)	6.24	5.64	ZPE (kcal/mol)	162.6	203
A (eV)	2.23	2.08	E (kcal/mol)	172.5	216.5
χ (eV)	4.235	3.86	C_v (cal/mol-K)	62.6	81.1
η (eV)	2.005	1.78	S (cal/mol-K)	125.7	154

Several thermodynamic parameters namely, zero-point energy (ZPE), thermal energy at room temperature (E), heat capacity (C_v), and entropy (S) for triclisine and rufescine are also calculated and listed in Table **3**. These parameters, which are related to one another via standard relations discussed in any textbook on thermodynamics, can be useful in determining the reaction paths of molecules.

IMERUBRINE

Molecular Structure

The optimized geometry of imerubrine is shown in Fig. (**4**). Its structure contains four planar, three hexaognal (R1, R2, and R4) as well as one pentagonal (R3) rings. The ring R1 is heterocyclic with the substitution of a nitrogen atom in the ring. The lower ring R2 is functionalized with three -OCH$_3$ groups whereas R4 is substituted with =O and -OCH$_3$. The imerubrine is similar to rufescine with a change in the position of the -OCH$_3$ group and the presence of an additional =O group.

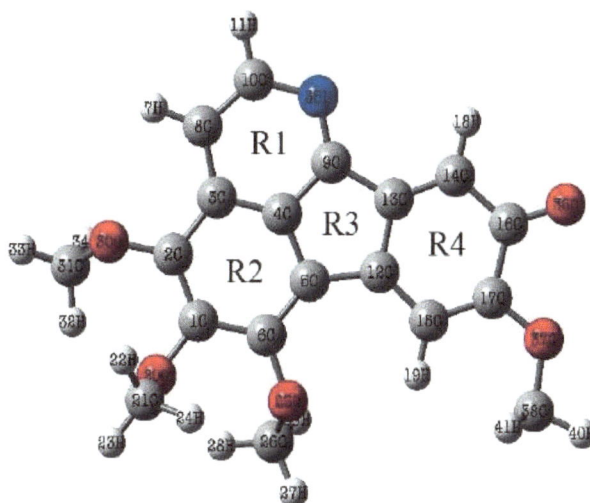

Fig. (4). Optimized geometry of imerubrine (reproduced from [20] with the permission of Springer).

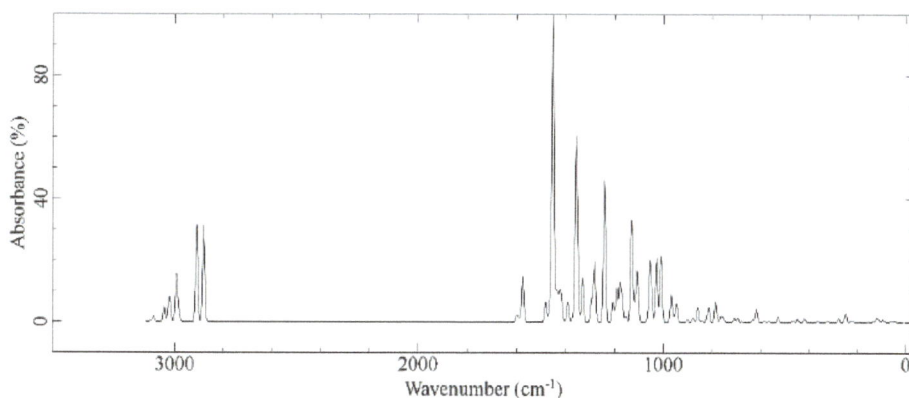

Fig. (5). Calculated IR spectrum of imerubrine (reproduced from [20] with the permission of Springer).

Vibrational Spectroscopic Analysis

Imerubrine ($C_{20}H_{17}NO_5$) possesses 123 normal modes of vibration (see Chapter 2). Herein, we present the assignments of normal modes of vibrations having wavenumber not below 400 cm^{-1} in Table **4**. We have assigned these modes by calculating their potential energy distribution (PED) with the help of the *VEDA* 4 program as described in the previous chapter. The calculated IR spectrum of imerubrine has been plotted in Fig. (**5**). We discuss the prominent vibrational modes of imerubrine as below.

C-H Vibrations

The C-H stretching vibration of aromatic compounds generally possess several bands in the region 3100-3000 cm^{-1} with weak intensities. Therefore, the vibrational modes calculated between 3108 and 3062 cm^{-1} correspond to the CH stretching of the rings. Similarly, the CH stretching vibrations related to the -OCH$_3$ groups are obtained in the region 3047-2914 cm^{-1}.

The in-plane CH bending modes can be usually found between 1300 and 1000 cm^{-1}. These vibrations have been obtained at 1270, 1265, and 1221 cm^{-1} in imerubrine. The out-of-plane CH bending modes become mixed with other modes, which can be observed in the range 1000-750 cm^{-1}. In imerubrine, these vibrational modes can be seen in the region 866-682 cm^{-1}.

Table 4. Infrared spectral analysis of imerubrine (reproduced from [20] with the permission of Springer).

Wavenumber (cm^{-1})		IR Intensity (a.u.)	Vibrational Assignments* (PED ≥ 15%)
Calculated	Scaled		
3221	3108	4.4	R4[v_{as}(C15-H19)(99)]
3216	3103	7.1	R1[v_{as}(C8-H7)(100)]
3210	3097	0.6	R4[v_{as}(C14-H18)(100)]
3174	3062	21.6	R1[v_s(C8-H7)(99)]
3158	3047	17.6	v_{as}(C38-H39)(99)
3155	3044	13	v_{as}(C31-H32)(90)
3153	3042	11.9	v_{as}(C26-H28)(92)
3148	3037	22.4	v_s(C14-H18)(97)
3120	3010	28.4	v_s(C31-H32)(92)
3118	3008	19.1	v_s(C26-H28)(86)
3108	2999	15.7	v_{as}(C21-H22)(100)
3090	2981	26.9	v_s(C38-H39)(99)
3034	2927	91.8	v_s(C31-C32)(83)
3033	2926	40	v_s(C26-H28)(71)
3031	2924	56.1	v_s(C21-H22)(90)
3020	2914	47.2	v_s(C38-H39)(100)
1660	1602	17.1	R2[v_{as}(C3-C4)(34)]
1654	1596	2.4	R2[v_s(C4-C3)(52)]
1639	1581	132.3	R2[v_{as}(C4-C3)(23)]
1625	1567	68.6	R2[v_s(C4-C3)(34)]
1574	1519	257.1	R4[v_{as}(C16-O36)(72)]
1532	1478	256.3	R2[v_s(C1-C2)(23)]+ω(H39-C38-H41)(15)
1527	1473	27.5	R2[v_{as}(C3-C2)(15)]

(Table 4) cont.....

1513	1460	156.1	R1[σ(H11-C10-N35)(17)]
1507	1454	60.8	τ(H28-C26-H27)(65)+ R2[$τ_i$(H28-C26-O25-C6)(15)]
1505	1452	18	τ(H28-C26-H27)(71)+ R2[$τ_i$(H28-C26-O25-C6)(17)]
1497	1444	40.2	σ(H22-C21-H24)(62)
1492	1439	39.6	σ(H39-C38-H41)(15)+ R4[$τ_i$(H39-C38-O37-C17)(17)]
1489	1437	66	σ(H22-C21-H24)(25)+ τ(H39-C38-H41)(20)
1488	1436	19.7	τ(H39-C38-H41)(70)+ R4[$τ_i$(H39-C38-O37-C17)(20)]
1485	1433	14.9	τ(H22-C21-H24)(69)+ R2[$τ_i$(H22-C21-O20-C1)(15)]
1484	1432	33.2	τ(H28-C26-H27)(24)+ ω(H34-C31-H33)(19)
1482	1430	4	τ(H22-C21-H24)(28)+ τ(H28-C26-H27)(22)
1478	1426	83.5	τ(H28-C26-H27)(22)+ σ(H34-C31-H33)(15)
1466	1414	4.3	τ(H28-C26-H27)(36)
1461	1410	49.1	ω(H39-C38-H41)(50)
1445	1394	65.3	ω(H22-C21-H24)(18)
1424	1374	199.1	R1[v_{as}(N35-C9)(16)]
1411	1361	138.4	R2[v_s(C2-C3)(14)]
1405	1356	195	R2[v_{as}(O20-C1)(15)]
1362	1314	30.3	R2[v_{as}(C2-C3)(15)+ v_{as}(O25-C6)(15)]
1358	1310	47.8	R2[v_{as}(C1-C2)(18)]
1316	1270	49.4	R2[v_s(C1-C2)(16)+σ(H18-C14-C16)(16)]
1311	1265	111.7	R1[σ(H11-C10-N35)(23)]
1266	1221	154.7	R4[σ(H18-C14-C16)(16)]
1251	1207	74.5	R2[v_s(O25-C6)(11)]
1217	1174	48.2	R2[$τ_i$(H28-C26-O25-C6)(21)]
1212	1169	23.6	R4[σ(H18-C14-C16)(46)]
1206	1163	1.7	R2[$τ_o$(H22-C21-O20-C1)(40)]
1203	1161	5	R2[$τ_i$(H28-C26-O25-C6)(31)]
1199	1157	5.4	R4[$τ_i$(H39-C38-O37-C17)(30)]
1176	1134	127.8	R4[$τ_o$(H39-C38-O37-C17)(13)]
1166	1125	0.2	R2[$τ_o$(H22-C21-O20-C1)(28)+ω(H22-C21-H24)(15)]
1165	1124	30	R2[$τ_i$(H28-C26-O25-C6)(34)]
1164	1123	28.8	R2[$τ_i$(H22-C21-O20-C1)(28)+ $τ_i$(H28-C26-O25-C6)(16)]
1119	1080	82.1	v_s(O20-C21)(18)
1104	1065	27.3	R1[σ(H7-C8-C10)(28)]
1096	1057	37.1	R2[v_s(C2-C3)(16)]+ v_s(O20-C21)(20)

(Table 4) cont....

1055	1018	41.9	υ_{as}(O37-C38)(18)+ υ_s(O20-C21)(27)
1044	1007	138.8	υ_s(O37-C38)(36)+ υ_s(O20-C21)(27)
1005	970	60.2	υ_s(O20-C21)(47)
992	957	8.9	υ_{as}(O20-C21)(31)
985	950	0.1	R2[τ_o(H7-C8-C3-C2)(85)]
939	906	5.3	R1[σ(C4-C9-N35)(16)]
916	884	7.1	υ_{as}(O20-C21)(20)
898	866	23	R4[τ_o(H18-C14-C17-C16)(56)]
868	837	12.5	R4[τi(H18-C14-C17-C16)(56)]
864	834	2.7	R2[σ(O20-C1-C2)(13)]
845	815	27.1	R2[τ_o(H7-C8-C3-C2)(79)]
795	767	6.7	R2[τ_o(O20-C6-C2-C1)(18)+ τ_o(C4-C8-C3-C2)(18)]
787	759	11.3	R2[τ_o(O20-C6-C2-C1)(15)+ τ_i(C3-C4-C9-N35)(20)]
773	746	18.8	R2[σ(O20-C1-C2)(17)]
760	733	6.7	R1[σ(C4-C9-N35)(10)]
757	730	0.5	R4[τ_o(H18-C14-C17-C16)(21)+ τ_o(O36-C14-C17-C16)(27)]
719	694	3.2	R2[τ_o(O25-C5-C1-C6)(15)+ τ_o(C12-C6-C4-C5)(18)]
707	682	1.6	R3[τ_o(C14-C9-C12-C13)(36)]
702	677	1.2	R1[σ(C4-C9-N35)(18)]
647	624	19.3	R2[ρ(C1-C2-C3)(14)]
631	609	1.6	R4[σ(O36-C14-C16)(13)]
590	569	0.2	R1[τ_i(C3-C4-C9-N35)(14)]
574	554	3.9	R2[τ_o(C1-C2-C3-C8)(12)]
499	481	0.6	R2[ρ(C4-C2-C3)(15)]
471	454	17.8	R2[σ(C1-C2-C3)(12)]
470	453	0.2	R4[τ_o(O36-C14-C17-C16)(25)+τ_o(O25-C6-C1-C5)(21)]
451	435	4.0	R4[σ(C38-O37-C17)(13)]

*Abbreviations: υ_{as}:asymmetric stretching, υ_s: symmetric stretching, σ: scissoring, ρ: rocking, τ_o: out-of-plane torsion, τ_i: in- plane torsion, R1, R2, R3, and R4: ring systems.

C-C Vibrations

Aromatic ring compounds possess CC stretching modes in the wavenumber range 1650-1200 cm^{-1}. These vibrational modes are greatly affected in the presence of functional groups. The wavenumbers of these bands are decided by the position of substitutions not usually by nature of the substitution. The C-C stretching modes are identified in the range 1602-1473 cm^{-1} in imerubrine. The CC bending modes have been obtained at lower wavenumbers, as included in Table **4**.

C-N Vibrations

The bands observed within the range 1400-1200 cm^{-1} can be attributed to the CN stretching for aromatic systems. These vibrations can not be identified easily because of the coupling of vibrational modes. The CN stretching in imerubrine has been obtained at 1374 cm^{-1}. The out-of-plane and in-plane C-N bending modes are also calculated, as listed in Table **4**.

C-O and C=O Vibrations

An intense vibrational peak corresponding to the C=O stretching has been identified at 1519 cm^{-1} in imerubrine. The peaks obtained between 1356 and 884 cm^{-1} correspond to the C-O stretching modes, which are shifted because of the lone-pair delocalization. These results are found to be in accordance with the value reported in literature. The other vibrational modes are also identified in the lower region (see Table **4**).

NMR Spectroscopic Analysis

^{1}H- and ^{13}C-NMR chemical shifts were obtained using the "gauge-independent atomic orbital" (GIAO) method) using B3PW91/6–311+G(d,p) level (see Chapter 2 for more details). The ^{1}H- and ^{13}C-NMR shielding values for imerubrine are shown in Fig. (**6**). The calculated ^{1}H and ^{13}C-NMR chemical shifts are collected in Table **5** along with corresponding experimental values as reported in the literature [17].

The chemical shifts (δ) of non-equivalent protons possess different values. This appears as a consequence of the strong deshielding because of the nonbonding electrons of the carbon atom. Hence, the calculated δ values of 7H (8.02 ppm), 11H (8.94 ppm), 18H (7.37 ppm), and 19H (7.64 ppm) have been found to be in accordance with the experimental values. The H atoms of the methyl group in imerubrine absorb far upfield but the corresponding δ values of 23H (4.14 ppm), 24H (3.97 ppm), and 27H (4.23 ppm), which are also in accordance with respective experimental values, 4.04, 4.00, and 4.12 ppm. The diamagnetic anisotropy results in these unusual shifts.

The peaks observed in the range between 100 and 150 ppm in the ^{13}C-NMR spectrum are due to aromatic carbon atoms. The chemical shifts of carbon atoms 5C, 13C, and 15C are obtained to be 128.3, 145.3, and 114.0 ppm, which is in good agreement with the corresponding experimental values. The chemical shifts of 9C and 10C are calculated at 163.0 and 152.0 ppm, respectively due to electron transfer from nitrogen atom. For the carbon of methyl group in imerubrine, chemical shifts are determined at 62.2 and 57.6 ppm. In general, the calculated values of ^{13}C-NMR chemical shifts agree with the corresponding experimental data.

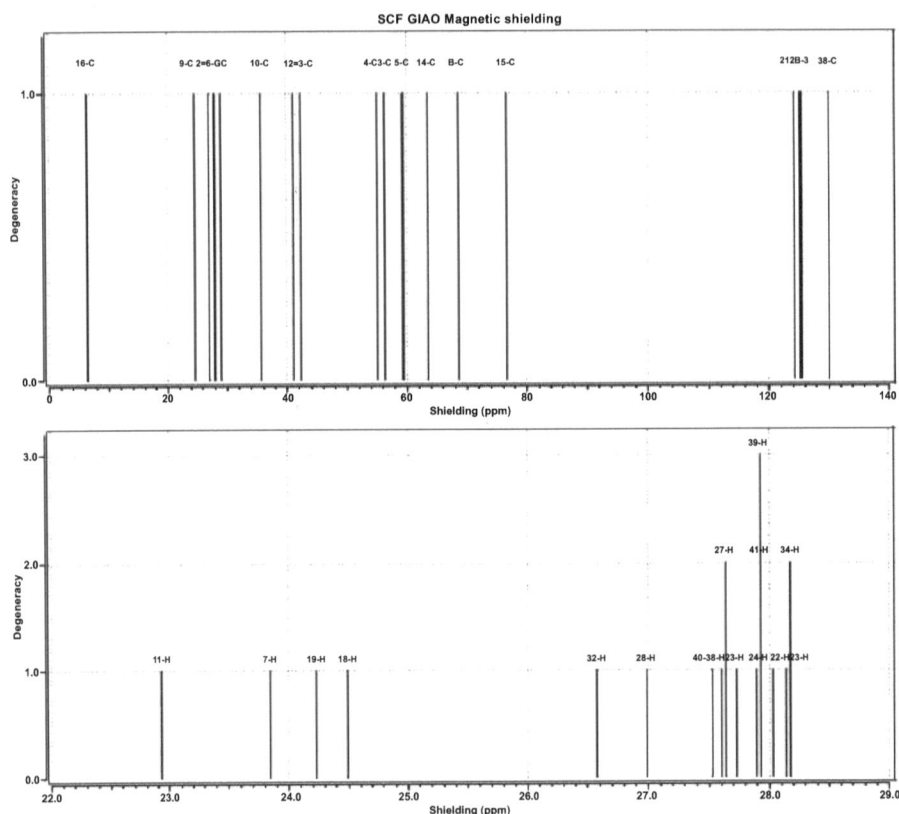

Fig. (6). Absolute C- and H-NMR shielding in imerubrine (reproduced from [20] with the permission of Springer).

Table 5. H- and C-NMR chemical shifts (δ, in ppm) and their assignments (s = singlet, d = doublet) (reproduced from [20] with the permission of Springer).

Atom	δ		Assignment	Atom	δ		Assignment
	Calc.	**Expt.**			**Calc.**	**Expt.**	
7H	8.02	8.05	[s, H(R1)]	5C	128.3	128.4	[s, C(R2)]
11H	8.94	8.67	[s, H(R1)]	9C	163	164.1	[s, C(R1)]
18H	7.37	6.86	[s, H(R4)]	10C	152	151.1	[s, C(R1)]
19H	7.64	7.74	[s, H(R4)]	13C	145.3	145.3	[s, C(R4)]
23H	4.14	4.04	[s, H(-OCH₃)]	15C	114	115	[s, C(R4)]
24H	3.97	4	[s, H(-OCH₃)]	31C	62.2	61.9	[s, C(-OCH₃)]
27H	4.23	4.12	[d, H(-OCH₃)]	38C	57.6	56.4	[s, C(-OCH₃)]

Chemical Reactivity

The chemical reactivity describes the tendency and preference of any system for a particular chemical interaction as mentioned earlier. The frontier orbitals, MESP and descriptors of to explore the reactivity of imerubrine is discussed as under.

HOMO, LUMO, and MESP Analyses

Fig. (7) plots the frontier orbitals surfaces of imerubrine. It can be seen that the HOMO is contributed by all rings, except the ring R1, unlike rufescine having HOMO consisted of all rings (see Fig. 3).

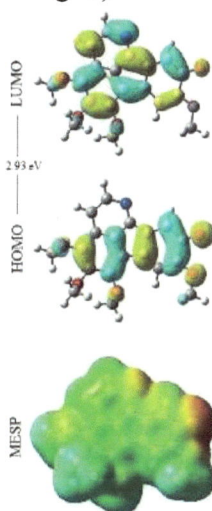

Fig. (7). The frontier orbitals and MESP surfaces of imerubrine (reproduced from [20] with the permission of Springer).

However, the LUMO of imerubrine is contributed by all four rings. Consequently, the transition from HOMO to LUMO leads to the transfer of electron density to the ring R1. This is because of the fact that the upper ring is substituted with an electronegative atom (N). The E_{gap} of 2.93 eV determines the strength of this charge-transfer interaction. Furthermore, this value is smaller than the E_{gap} of rufescine (3.56 eV), which indicates that imerubrine is chemically more reactive than rufescine.

The molecular electrostatic potential (MESP) is very important in determining the susceptibility of molecule for electrophilic (electronegative region) and nucleophilic (electropositive region) reactions [24]. The MESP also plays a key role in understanding the phenomena of the molecular recognition, *e.g.*, in drug-receptor interactions, due to the fact two species face each other through their potentials [25]. Fig. (7) also displays the MESP surface of imerubrine in a color-grading scheme. The color grade for imerubrine ranges between +0.07022 a.u. (blue) and -0.07022 a.u. (red), where blue and red represent the highly electropositive and electronegative regions, respectively as mentioned earlier. The MESP of imerubrine shows that the oxygen attached to the ring R4 becomes the highly electronegative, *i.e.*, very prone to electrophilic substitution. In rufescine, however, the nitrogen attached to the ring R1 is highly electronegative, *i.e.*, very likely to act as an electron donor (see Fig. **3**).

Reactivity Descriptors

Various electronic parameters can be used to further describe the reactive nature of imerubrine. These reactivity descriptors can be found in Table **6**. It is evident that imerubrine possesses smaller I and χ values but larger A and η values as compared to rufescine (Table **3**). This further suggests that imerubrine is chemically softer, *i.e.*, more reactive as compared to rufescine.

Table 6. Various reactivity descriptors of imerubrine in eV (reproduced from [20] with the permission of Springer).

Descriptor	Value
I	5.45
A	2.52
χ	3.985
ω	1.465

CONCLUDING REMARKS

We have discussed the DFT results of triclisine and rufescine using the B3PW91/6-311+G(d,p) method. Normal mode analyses have been performed, and complete assignments are given to the prominent modes of vibrations. The calculated infrared spectra of both molecules have been found to be in good agreement with corresponding experimental FT-IR spectra. The chemical reactivity of triclisine and rufescine has been discussed by frontier orbitals and MESP surfaces along with various electronic and thermodynamic parameters.

We have also discussed the results of imerubrine. The assignments of all vibrational modes > 400 cm^{-1} have been offered on the basis of potential energy distribution values. The H- and C-NMR chemical shifts of imerubrine have been obtained which agreed well with the corresponding experimental values. The transition from HOMO to LUMO leads to the transfer of electron density to heterocyclic ring in imerubrine and the smaller energy gap along with reactivity parameters suggest chemically more reactive behavior.

CONSENT FOR PUBLICATION

Not applicable.

CONFLICT OF INTEREST

The authors declare no conflict of interest, financial or otherwise.

ACKNOWLEDGEMENTS

Declared none.

REFERENCES

[1]   Rotard, W.; Mailahn, W. Gas chromatographic-mass spectrometric analysis of creosotes extracted from wooden sleepers installed in playgrounds. *Anal. Chem.,* **1987**, *59*(1), 65-69.
        [http://dx.doi.org/10.1021/ac00128a014] [PMID: 3826635]

[2]   Chen, H.Y.; Preston, M.R. Azaarenes in the aerosol of an urban atmosphere. *Environ. Sci. Technol.,* **1998**, *32*, 577-583.
        [http://dx.doi.org/10.1021/es970033n]

[3]   Hu, R.J.; Liu, H.X.; Zhang, R.S.; Xue, C.X.; Yao, X.J.; Liu, M.C.; Hu, Z.D.; Fan, B.T. QSPR prediction of GC retention indices for nitrogen-containing polycyclic aromatic compounds from heuristically computed molecular descriptors. *Talanta,* **2005**, *68*(1), 31-39.
        [http://dx.doi.org/10.1016/j.talanta.2005.04.034] [PMID: 18970281]

[4]   Barron, M.G.; Heintz, R.; Rice, S.D. Relative potency of PAHs and heterocycles as aryl hydrocarbon receptor agonists in fish. *Mar. Environ. Res.,* **2004**, *58*(2-5), 95-100.
        [http://dx.doi.org/10.1016/j.marenvres.2004.03.001] [PMID: 15178019]

[5]   Yan, M.H.; Cheng, P.; Jiang, Z.Y.; Ma, Y.B.; Zhang, X.M.; Zhang, F.X.; Yang, L.M.; Zheng, Y.T.; Chen, J.J. Periglaucines A-D, anti-HBV and -HIV-1 alkaloids from Pericampylus glaucus. *J. Nat. Prod.,* **2008**, *71*(5), 760-763.
        [http://dx.doi.org/10.1021/np070479+] [PMID: 18396905]

[6]   Morita, H.; Matsumoto, K.; Takeya, K.; Itokawa, H. Azafluoranthene Alkaloids from Cissampelos pareira. *Chem. Pharm. Bull. (Tokyo),* **1993**, *41*, 1307-1308.
        [http://dx.doi.org/10.1248/cpb.41.1307]

[7]   Swaffar, D.S.; Holley, C.J.; Fitch, R.W.; Elkin, K.R.; Zhang, C.; Sturgill, J.P.; Menachery, M.D. Phytochemical investigation and *in vitro* cytotoxic evaluation of alkaloids from *Abuta rufescens.* *Planta Med.,* **2012**, *78*(3), 230-232.
        [http://dx.doi.org/10.1055/s-0031-1280383] [PMID: 22109836]

[8]   Khan, S.I.; Nimrod, A.C.; Mehrpooya, M.; Nitiss, J.L.; Walker, L.A.; Clark, A.M. Antifungal activity of eupolauridine and its action on DNA topoisomerases. *Antimicrob. Agents Chemother.,* **2002**, *46*(6), 1785-1792.
        [http://dx.doi.org/10.1128/AAC.46.6.1785-1792.2002] [PMID: 12019091]

[9]   Gasiorski, P.; Danel, K.S.; Matusiewicz, M.; Uchacz, T.; Vlokh, R.; Kityk, A.V. Synthesis and spectroscopic study of several novel annulated azulene and azafluoranthene based derivatives. *J. Fluoresc.,* **2011**, *21*(1), 443-451.
        [http://dx.doi.org/10.1007/s10895-010-0722-1] [PMID: 20886270]

[10]   Gasiorski, P.; Danel, K.S.; Matusiewicz, M.; Uchacz, T.; Kuźnik, W.; Kityk, A.V. DFT/TDDFT study on the electronic structure and spectral properties of diphenyl azafluoranthene derivative. *J. Fluoresc.,*

2012, *22*(1), 81-91.
[http://dx.doi.org/10.1007/s10895-011-0932-1] [PMID: 21853258]

[11]   Danel, K.S.; Gasiorski, P.; Matusiewicz, M.; Całus, S.; Uchacz, T.; Kityk, A.V. UV-vis spectroscopy and semiempirical quantum chemical studies on methyl derivatives of annulated analogues of azafluoranthene and azulene dyes. *Spectrochim. Acta A Mol. Biomol. Spectrosc.*, **2010**, *77*(1), 16-23.
[http://dx.doi.org/10.1016/j.saa.2010.04.006] [PMID: 20510645]

[12]   Calus, S.; Danel, K.S.; Uchacz, T.; Kityk, A.V. Optical absorption and fluorescence spectra of novel annulated analogues of azafluoranthene and azulene dyes. *Mater. Chem. Phys.*, **2010**, *121*, 477-483.
[http://dx.doi.org/10.1016/j.matchemphys.2010.02.010]

[13]   Ponnala, S.; Harding, W.W. A new route to azafluoranthene natural products via direct arylation. *Eur. J. Org. Chem.*, **2013**, *3013*(6), 1107-1115.
[http://dx.doi.org/10.1002/ejoc.201201190] [PMID: 23503080]

[14]   Silveira, C.C.; Larghi, E.L.; Mendes, S.R.; Bracca, A.B.J.; Rinaldi, F.; Kaufman, T.S. Electrocyclization☐Mediated Approach to 2☐Methyltriclisine, an Unnatural Analog of the Azafluoranthene Alkaloid Triclisine. *Eur. J. Org. Chem.*, **2009**, *2*, 4637-4645.
[http://dx.doi.org/10.1002/ejoc.200900673]

[15]   Srivastava, A.K.; Pandey, A.K.; Jain, S.; Misra, N. FT-IR spectroscopy, intra-molecular C-H☐O interactions, HOMO, LUMO, MESP analysis and biological activity of two natural products, triclisine and rufescine: DFT and QTAIM approaches. *Spectrochim. Acta A Mol. Biomol. Spectrosc.*, **2015**, *136*(Pt B), 682-689.
[http://dx.doi.org/10.1016/j.saa.2014.09.082] [PMID: 25315865]

[16]   Buck, K.T. *Alkaloids, Academic Press,* **1984**, *23*, 301.

[17]   Zhao, B.; Snieckus, V. Integrated Aromatic Metalation-Cross Coupling Methodologies. A Concise Synthesis of the Azafluoranthene Alkaloid Imeluteine. *Tetrahedron Lett.*, **1984**, *32*, 5277-5278.
[http://dx.doi.org/10.1016/S0040-4039(00)92363-3]

[18]   Boger, D.L.; Takahashi, K. Total Synthesis of granditropone, grandirubrine, imerubrine and isoimerubrine. *J. Am. Chem. Soc.*, **1995**, *117*, 12452-12459.
[http://dx.doi.org/10.1021/ja00155a009]

[19]   Lee, J.C.; Cha, J.K. Total synthesis of tropoloisoquinolines: imerubrine, isoimerubrine, and grandirubrine. *J. Am. Chem. Soc.*, **2001**, *123*(14), 3243-3246.
[http://dx.doi.org/10.1021/ja0101072] [PMID: 11457059]

[20]   Srivastava, A.K.; Kumar, A.; Pandey, S.K.; Misra, N. Spectroscopic analyses, intra-molecular interaction, chemical reactivity and molecular docking of imerubrine into bradykinin receptor. *Med. Chem. Res.*, **2016**, *25*, 2832-2841.
[http://dx.doi.org/10.1007/s00044-016-1710-z]

[21]   Sklenar, H.; Jager, J. Molecular structure-biological activity relationships on the basis of quantum-chemical calculations. *Int. J. Quantum Chem.*, **1979**, *16*, 467-484.
[http://dx.doi.org/10.1002/qua.560160306]

[22]   Murray, J.S.; Sen, K. *Molecular Electrostatic Potentials: Concepts and Applications*; Elsevier: Amsterdam, Netherlands, **1996**.

[23]   Sponer, J.; Hobza, P. DNA base amino groups and their role in molecular interactions: Ab initio and preliminary density functional theory calculations. *Int. J. Quantum Chem.*, **1996**, *57*, 959-970.
[http://dx.doi.org/10.1002/(SICI)1097-461X(1996)57:5<959::AID-QUA16>3.0.CO;2-S]

[24]   Luqul, F.J.; Lopez, J.M.; Orozco, M. Perspective on Electrostatic interactions of a solute with a continuum, A direct utilization of ab initio molecular potentials for the prevision of solvent effects. *Theor. Chem. Acc.*, **2000**, *103*, 343.
[http://dx.doi.org/10.1007/s002149900013]

[25]　Scrocco, E.; Tomasi, J. The electrostatic molecular potential as a tool for the interpretation of molecular properties. *Top. Curr. Chem.,* **1973**, *42*, 95-170.
[http://dx.doi.org/10.1007/3-540-06399-4_6]

CHAPTER 6

A Comprehensive DFT Study on a Thione Compound and its Tautomer

Abstract: In this chapter, we present an exhaustive study on a synthetic compound, 4-Amino-3-(4-hydroxybenzyl)-1H-1,2,4-triazole-5(4H)-thione (4AHT) and its tautomer employing both experimental and theoretical methods. 4AHT was synthesized by the reaction of 4-hydroxyphenylacetic acid and thiocarbohydrazide. The derivatives of 1,2,4-triazole are known to display antiviral, antidepressant, antimicrobial and anti-inflammatory activities. A DFT study on thione and thiol tautomers of 4AHT has been performed using the B3LYP/6-31+G(d,p) level. The calculated geometrical parameters are obtained to be in compliance with corresponding crystallographic values. The FT-IR spectrum of 4AHT obtained by the KBr disc technique has been explained by assigning the normal modes based on their potential energy distributions. The UV-visible spectrum of the title compound has been discussed by calculating the electronic transitions in several excited states using the TD-DFT method. The ^{1}H-NMR spectrum has also been studied by the GIAO method, which provides a good linear correlation between calculated and experimental chemical shifts. These spectroscopic results propose the dominance of the thione form of 4AHT in the solid-state and NH-SH tautomerism in the liquid form. This chapter is designed to provide a complete picture of the DFT based studies on molecular systems.

Keywords: B3LYP, DFT, Electronic parameter, Electronic transition, Experiment, FT-IR, HOMO, LUMO, MESP, NH-SH tautomerism, NMR, PES, Reaction, Synthesis, Tautomer, TD-DFT, Thermodynamics, Thiol, Thione, UV-visible.

INTRODUCTION

The well known heterocyclic triazoles have been recently in focus due to their adaptability for the synthesis of various heterocyclic systems. There exist two tautomers of triazoles, namely, the 1,2,3-triazole, symbolized as 1H and the 1,2,4-triazole, symbolized as 4H. The planar 4H rings are aromatic with 6π-electron-systems, which are very important in chemical science [1, 2]. The derivatives 1,2,4-triazole exhibit several biological activities, and hence, many 4H derivatives along with their N-bridged compounds were paid attention to in the past [3, 4]. The 1,2,4-triazole derivatives have been found to show anti-inflammatory [5],

Ambrish Kumar Srivastava and Neeraj Misra

antiviral [6], antimicrobial [7], and antidepressant [8] activities. Therefore, the synthesis and extensive study of these heterocyclic systems are of immense importance.

Schiff bases [9, 10] and related heterocycles including triazoles [11 - 13] become potent corrosion inhibitors. Their chemisorption primarily takes place due to the −C=N− group interaction [10] and electronegative atoms such as N, O, S, *etc.* in the Schiff bases make them efficient corrosion inhibitors. Several compounds based on triazoles are reported to be good corrosion inhibitors for steel in acid media [14 - 17]. The synthesis of a few 1,2,4-triazole-5(4H)-thione compounds have been reported by Salgın-Gökşen *et al.* [18] along with their antimicrobial, analgesic, and anti-inflammatory activities. In this chapter, a comprehensive experimental and theoretical study of 4-Amino-3-(4-hydroxybenzyl)-1H-1,2,4-triazole-5(4H)-thione (4AHT) is presented, which was performed by us [19]. The 4AHT compound was synthesized by fusing a well-triturated mixture of 4-hydroxyphenylacetic acid ($C_8H_8O_3$) and thiocarbohydrazide (CH_6N_4S) in a round bottom flask for one hour on an oil bath at 413 K and subsequently cooling to room temperature [20]. The compound, thus obtained, was dried and recrystallized using methanol, yielding colourless block-like crystals, being triclinic with a P_1 symmetry group.

The reaction thermodynamics for the synthesis and spectroscopic analyses are performed using the density functional theory (DFT) based method. The chemical reactivity of 4AHT is studied with the help of molecular electrostatic potential (MESP) surface and several descriptors. The biological activities of 4AHT are also anticipated. We are of the firm view that these results will be useful in the future works on the 1,2,4-triazole-5(4H)-thione and related systems.

METHODS

Experimental

The sample has been estimated as 100% pure using the HPLC method. The Shimadzu 8400S spectrometer with the resolution of ±4 cm^{-1} was employed to record the FT-IR spectrum of sample in the wavenumber range 4000-400 cm^{-1} using the KBr pellet (see Chapter 4). The Bruker 400 MHz NMR spectrometer was used to record the ^{1}H-NMR spectrum of 4AHT with dimethyl sulfoxide (DMSO) solvent. The Shimadzu UV-2550 double-beam spectrophotometer having 1 cm matched quartz cell to measure the absorbance was utilized to record the UV-visible spectrum of 4AHT with ethyl alcohol (ethanol) solvent.

Computational

The optimization of the structure of 4AHT was performed with no constraints on symmetry in the potential energy surface (PES) with DFT at the B3LYP [21, 22] functional and a 6-31+G(d,p) basis set. The UV-visible spectrum was replicated by applying the time-dependent DFT (TD-DFT) approach using the equilibrium structure of 4AHT at the B3LYP level. The gauge independent atomic orbital (GIAO) method was used to calculate the ^{1}H-NMR chemical shifts at the same level. The polarizable continuum model devised by Tomasi *et al.* [23] has been exploited to incorporate the solvent effect during spectroscopic calculations. These computations were carried out by the *Gaussian 09* software [24].

RESULTS AND DISCUSSION

Synthesis and Thermodynamics

Fig. (1) displays the schematic of the reaction for the synthesis of 4AHT. The thermodynamics of this reaction can be analyzed by calculating various thermal parameters of reactants and products using room temperature. Table 1 lists the room temperature values of total electronic energy (E), thermal Enthalpy (H), Gibbs' free energy (G), and entropy (S) for these reactants ($C_8H_8O_3$, CH_6N_4S) and products ($C_9H_{10}N_4OS$, $2H_2O$).

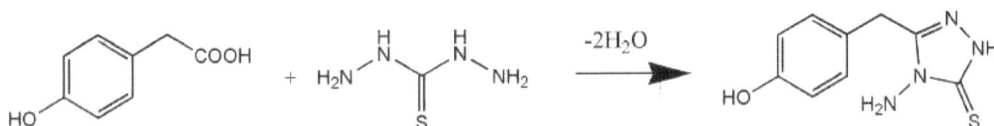

Fig. (1). Synthesis scheme of 4AHT [20].

One can see that the change in enthalpy change ($\Delta H_{\text{Reaction}}$), Gibbs free energy ($\Delta G_{\text{Reaction}}$), and entropy ($\Delta S_{\text{Reaction}}$) for this reaction are calculated to be -0.3010 a.u., -0.0415 a.u. and 24.8 cal/mol.K, respectively (1 a.u. = 627.5 kcal/mol). Thus, $\Delta H_{\text{Reaction}}$ and $\Delta G_{\text{Reaction}}$ possess negative values but the value of $\Delta S_{\text{Reaction}}$ is positive. This clearly reveals that the synthesis of 4AHT is favourable at room temperature due to the spontaneous and exothermic nature of the reaction.

Since both H and S favour the reaction path and so does G because $\Delta G = \Delta H - T \Delta S$. The equilibrium constant (K_{eq}) and $\Delta G_{\text{Reaction}}$ at temperature (T) are related as $K_{eq} = e^{-\Delta G/RT}$, which decides the nature (direction) of a chemical reaction. The K_{eq} value for the above mentioned reaction is found to be 1.285×10^{19} at R = 1.987×10^{-3} kcal/mol.K and $T = 298.15$ K. This further suggests that the reaction proceeds in a forward direction at room temperature as $K_{eq} > 1$. Thus, the thermodynamics of reaction supports the synthesis of 4AHT at room temperature.

Table 1. Thermodynamics of reaction at the room temperature (reproduced from [19] with the permission of Elsevier).

Systems	E	H	G	S
	(a.u.)	(a.u.)	(a.u.)	(cal/mol-K)
$C_9H_{10}N_4OS$	-141.2206	-141.2061	-1041.263	119.9
$C_8H_8O_3$	-535.2458	-535.235	-535.2833	101.621
CH_6N_4S	-658.7673	-658.7588	-658.799	84.703
$2H_2O$	-152.8254	-152.8178	-152.8608	90.2
Change ($\Delta_{Reaction}$)	-	-0.0301	-0.0415	24.776

The optimized geometry of 4AHT is displayed in Fig. (**2**). One can see that the phenyl ring is approximately normal to the triazole ring having the C3-C21-C13 bond angle of 113.5^0, in accordance with the experimental value of 113^0. The calculated structural parameters, *i.e.*, bond-lengths, bond-angles, and dihedrals of 4AHT, along with experimentally determined parameters using the X-ray crystallography can be found in Table **2**. Fig. (**3**) shows the correlation graph of calculated and experimental bond lengths and bond angles. It can be seen that there is a good correlation between both parameters having the linear correlation coefficient (R^2) of greater than 0.9. To confirm that the equilibrium structure belongs to some minimum in the PES, we carried out the scan of PES of 4AHT with respect to dihedral angles C3-C21-C13-N16 and C5-C6-O24-H25. The PES scan curves of 4AHT are shown in Fig. (**4**), which suggests the geometry of 4AHT (Fig. **2**) to be a true minimum.

Although 4AHT is found in thione form in the solid-state, it shows NH-SH tautomerism in a liquid state. Hence, we have also studied its thiol tautomer, whose optimized structure is also included in Fig. (**2**). Romani and Brandan [25] have already reported this tautomerism in another thione compound, namely, 6-Nitro-1,3-benzothiazole-2(3H)-thione.

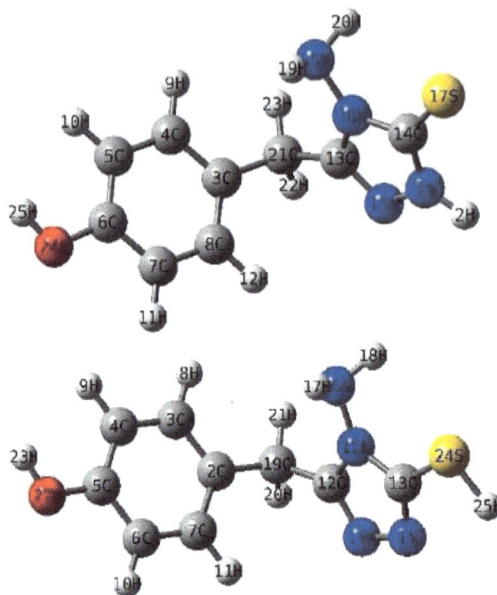

Fig. (2). Optimized structures of 4AHT in thiol (upper) and thiol (lower) forms (reproduced from [19] with the permission of Elsevier).

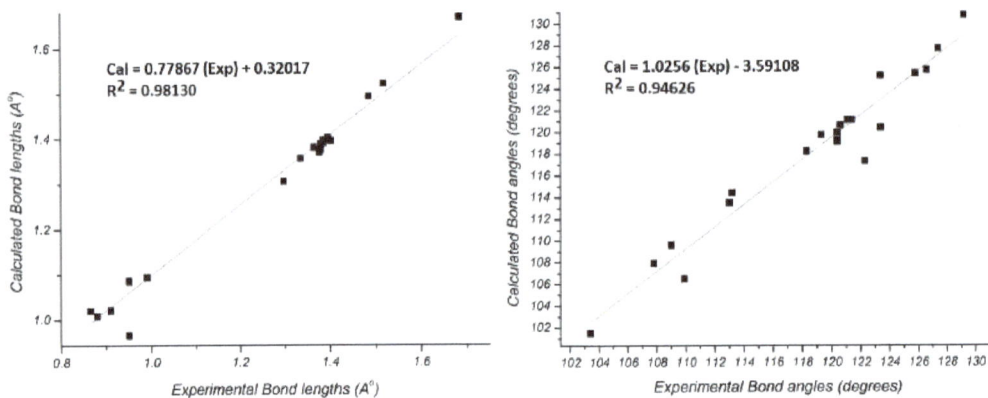

Fig. (3). Linear correlation graphs of calculated and experimental parameters (reproduced from [19] with the permission of Elsevier).

Table 2. Calculated geometrical and experimental crystallographic parameters (in Å, 0) of 4AHT (reproduced from [19] with the permission of Elsevier).

Parameter	Calc.	Expt.	Parameter	Calc.	Expt.
C3-C4	1.398	1.389	C4-C5	1.397	1.390
C5-C6	1.397	1.384	C6-C7	1.399	1.396
C7-C8	1.392	1.385	C3-C8	1.403	1.395

(Table 2) cont.....

C3-C21	1.523	1.517	C13-C21	1.495	1.484
C4-H9	1.086	0.950	C5-H10	1.087	0.950
C7-H11	1.085	0.950	C8-H12	1.087	0.950
C21-H22	1.094	0.990	C21-H23	1.094	0.990
C13-N16	1.306	1.297	C13-N15	1.381	1.364
C14-N1	1.357	1.334	C14-N15	1.388	1.380
C6-O24	1.370	1.376	H2-N1	1.008	0.880
H19-N18	1.020	0.865	H20-N18	1.021	0.910
N1-N15	1.376	1.379	N18-N16	1.396	1.403
C14-S17	1.670	1.683	H25-O24	0.966	0.950
H2-N1-C14	125.2	123.4	H2-N1-N15	120.4	123.4
C14-N1-N15	114.4	113.2	C4-C3-C8	118.2	118.3
C4-C3-C21	121.1	121.1	C8-C3-C21	120.6	120.6
C3-C4-C5	121.1	121.4	C3-C4-H9	119.7	119.3
C5-C6-C7	119.9	120.4	C7-C6-O24	117.3	122.3
C6-C7-H11	119.1	120.4	C7-C8-H12	119.2	120.4
N15-C13-C21	125.4	125.8	N1-C14-N16	101.5	103.4
N16-C14-S17	127.7	127.4	N1-C14-S17	130.8	129.2
C14-N16-N18	125.7	126.6	H19-N18-H20	106.5	109.9
C3-C21-H23	109.6	109.0	H22-C21-H23	107.9	107.8
C3-C21-C13	113.5	113.0	-	-	-
N15-N1-C14-S17	-178.5	-178.7	N15-N1-C14-N16	0.3	0.8
C14-N1-N15-C13	0.1	0.8	C8-C3-C4-C5	-0.2	-0.7
C21-C3-C4-C5	-179.1	-178.2	C4-C3-C8-C7	0.1	-0.9
C3-C4-C5-C6	0.04	0.7	O24-C6-C7-C8	-179.9	-176.7
S17-C14-N16-N18	1.5	-4.9	S17-C14-N16-C13	178.3	178.6
N1-C14-N16-N18	-177.4	177.5	N16-C13-C21-C3	-74.6	83.5
N15-C13-C21-C3	103	-94.8	C21-C13-N16-C14	178.5	-179.1

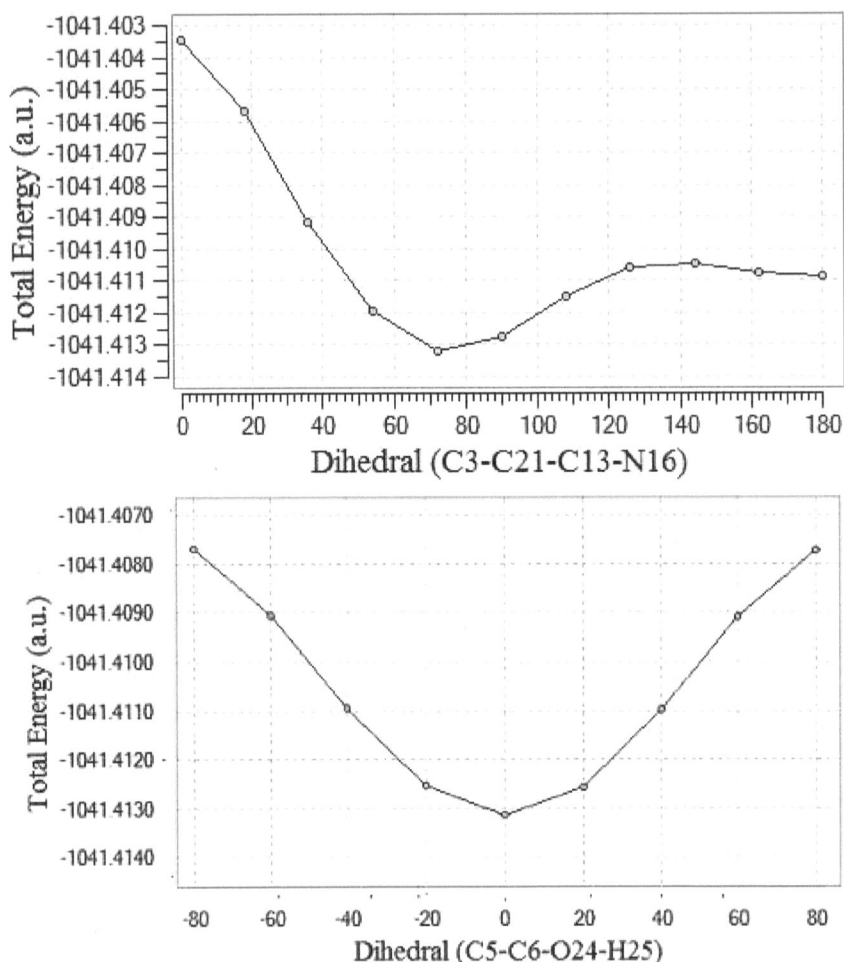

Fig. (4). PES scan curves of 4AHT (reproduced from [19] with the permission of Elsevier).

Infrared Spectroscopic Analysis

Table **3** lists the calculated and scaled wavenumbers of 4AHT, their assignments, IR intensities, and FT-IR wavenumbers. Considering the anharmonicity in vibrations and the deficiencies in the basis set employed, the B3LYP/6-31+G(d,p) calculated wavenumbers are scaled by scale factor ($v_{scaled} = 0.9648 v_{calculated}$) [26] as well as scaling equation ($v_{scaled} = 22.1 + 0.9543 v_{calculated}$) [27] (see Chapter 2 for elaborated discussion). It is evident from Table **3** that the scaling equation works good in the higher wavenumber region (> 1800 cm^{-1}) and scale factor does well in the lower wavenumber region (< 1800 cm^{-1}). This is why the scaling equation is preferred for scaling the normal modes for the higher wavenumber region in which the stretching vibrations mostly occur. The assignments of normal modes are performed by calculating their potential energy distribution (PED) using the

VEDA 4 program [28, 29] and graphical animations of the *GaussView 5.0* program [30]. The calculated vibrational (infrared) spectra of both tautomers along with FT-IR spectrum of 4AHT are shown in Fig. (**5**).

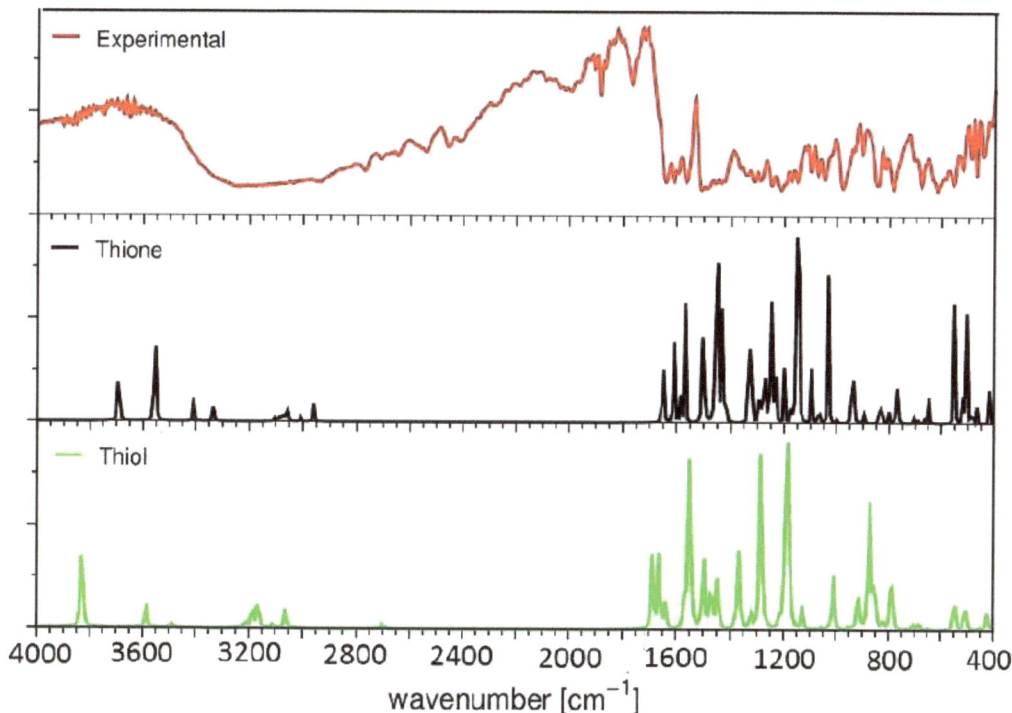

Fig. (5). Experimental FT-IR spectrum of 4AHT and calculated IR spectra of thione and thiol tautomers (reproduced from [19] with the permission of Elsevier).

Table 3. Calculated and scaled wavenumbers, infrared intensity and their assignments. The corresponding FT-IR wavenumbers and intensities are also included for comparison (reproduced from [19] with the permission of Elsevier).

Calc. Wavenumber	Scaled Wavenumber (cm^{-1})		IR Int.	FT-IR Value	Vibrational Assignment*
(cm^{-1})	ScaleFact.	ScalingEqn.	(a.u.)	(cm^{-1})	(PED ≥ 10%)
3826	3691	3674	69.4	-	υ(OH) (100)
3683	3554	3537	110.9	-	υ(1N-2H) (99)
3529	3405	3390	12.8	-	υ_{as}(NH$_2$) (99)
3454	3332	3319	11.8	3330 (s)	υ_s(NH$_2$) (98)
3212	3099	3088	4.2	3134 (s)	R[υ_{as}(4C-9H) (34)+ υ_s(5C-10H) (42)]
3194	3082	3071	8.7	-	R[υ_s(7C-11H) (42)+ υ_s(8C-12H) (38)]

(Table 3) cont.....

3184	3072	3061	4	-	R[v_{as}(7C-11H) (42)+ v_{as}(8C-12H) (33)]
3168	3057	3046	18.7	-	R[v_s(4C-9H) (37)+ v_{as}(5C-10H) (43)]
3116	3007	2996	2.1	-	v_{as}(CH$_2$) (99)
3067	2959	2949	10.9	2935 (m)	v_s(CH$_2$) (99)
1705	1645	1650	66.4	1639 (m)	σ(NH$_2$) (81)
1663	1604	1609	55	-	R[v_{as}(4C-3C) (50)]
1638	1580	1586	17.5	1589 (m)	R[v_{as}(4C-3C) (50)]
1620	1562	1568	37	1564 (m)	R[v_s(15N-13C) (61)]
1550	1496	1502	113.1	1507 (s)	R[σ(9H-4C-5C) (31)]
1499	1446	1453	294.7	1447 (m)	R[v_s(1N-14C) (40)+σ(2H-1N-15N) (19)]
1478	1426	1433	34.2	-	R[v_{as}(1N-14C) (10)]
1470	1418	1425	3.2	-	σ(21C-22H-23H) (50)
1465	1413	1421	18.4	-	σ(21C-22H-23H)(27)+R[τ_o(22H-21C--C-8C)(15)]
1377	1328	1336	25.8	-	R[ρ(2H-1N-15N) (23)]
1369	1321	1329	40.6	-	R[v_{as}(4C-3C)(14)]
1362	1314	1323	5.4	-	R[τ_t(22H-21C-3C-8C)(19)]
1334	1287	1296	34.5	1290 (m)	ω(22H-21C-3C)(10)
1316	1269	1278	74.7	-	R[σ(14C-1N-15N)(16)]+ τ_o(22H-21C--C-8C)(22)
1290	1245	1253	122.9	1250 (s)	R[v_{as}(4C-3C)(16)]+R[v_s(24O-6C)(48)]
1273	1228	1238	51.5	-	R[v_s(1N-14C)(32)+ σ(2H-1N-15N)(17)]
1237	1194	1203	17.2	1210 (s)	τ(22H-21C-3C)(16)
1214	1171	1181	6	-	R[v_s(4C-3C)(22)+ v_s(3C-21C)(34)]
1199	1157	1167	14.5	-	R[v_s(4C-3C)(17)+σ(9H-4C-5C)(74)]
1186	1144	1154	162.5	1150 (m)	R[σ(25H-24O-6C)(43)]
1179	1137	1147	40.1	-	R[σ(25H-24O-6C)(16)+σ(22H-21C-3C)(16)]
1129	1089	1099	21.3	1100 (m)	R[v_{as}(4C-3C)(16)

(Table 3) cont.....

1100	1061	1072	16.4	1074 (m)	R[v_s(1N-15N)(53)]
1062	1025	1036	59.2	-	R[τ_o(2H-1N-15N-13C)(31)
1029	992	1004	1.6	-	R[σ(4C-3C-21C)(52)]
975	940	953	0.14	-	R[τ_i(9H-4C-5C-6C)(81)]
948	914	927	0.1	-	R[τ_i(9H-4C-5C-6C)(59)]
921	888	901	3.9	900 (w)	R[τ_i(22H-21C-3C-8C)(26)]
860	830	843	13.6	846 (m)	R[τ_i(9H-4C-5C-6C)(15)]
852	822	835	22.1	-	R[τ_i(9H-4C-5C-6C)(21)]
791	763	778	30.5	785 (s)	R[σ(3C-21C-13C)(13)]
786	758	772	10.2	-	R[v_{as}(3C-21C)(26)]
723	697	712	1.7	-	R[v_{as}(18N-16N)(36)+σ(14C-1N-15N)(16)]
705	680	695	1.9	-	R[τ_o(4C-3C-21C-13C)(30)]
678	654	669	5.5	675 (m)	R[τ_o(1N-15N-13C-16N)(27)]
651	628	643	1.5	-	R[σ(4C-3C-21C)(62)]
564	544	560	47.2	552 (m)	R[v_s(17S-14C)(13)]
494	477	494	13.2	495 (w)	R[τ_i(1N-15N-13C-16N)(14)]
474	457	475	5.8	449 (w)	R[v_s(17S-14C)(18)+σ(17S-14C-1N)(16)]
423	408	426	11.2	-	R[σ(4C-3C-21C)(27)+σ(24O-6C-5C)(60)]
421	407	425	0.8	-	R[τ_o(4C-3C-21C-13C)(57)]
336	324	344	106.6		R[τ_i(25H-24O-6C-7C)(93)]
300	289	309	3.2		R[σ(4C-3C-21C)(14)]
143	138	159	6.3		R[σ(3C-21C-13C)(15)]
109	105	127	2		R[τ_o(14C-1N-15N-13C)(19)]

*Abbreviations: R: Ring, v_s: symmetric stretching, v_a: antisymmetric stretching, σ: scissoring, ω: wagging, ρ: rocking, τ: twisting, τ_i: in-plane bending, τ_o: out-of-plane bending, s: strong, m: medium, w: weak.

Ring Vibrations

The CH stretching modes in phenyl rings are identified in the wavenumber range 3100-3000 cm^{-1} which is the characteristic region for the CH stretching vibrations [31 - 33]. The CH stretching modes in 4AHT are obtained in the region between 3088 and 2949 cm^{-1}. These vibrations, being purely stretching, possess PED exceeding 90%, which are in accordance with the experimental peaks determined at 3134 cm⁻¹ and 2935 cm⁻¹. These results also tune with the literature values for aromatic CH stretching, *i.e.*, 3086-3026 cm⁻¹ [34] and 3074-3037 cm⁻¹ [31] in our previous studies. The CH scissoring coming from in-plane and out-of-plane CH twisting modes associated with the ring are calculated beneath 1500 cm⁻¹ with weak to medium intensities. The CC stretching modes in phenyl rings are found to be in the range 1604-1089 cm⁻¹, which is also in accordance with the values given in the literature [35, 36]. In the lower wavenumber region, a few more CC modes, coupled with other vibrations are also present. Usually, CC vibrations possess relatively higher intensities than CH modes.

The stretching modes obtained within the range 1562-1228 cm⁻¹ are associated with the triazole ring, which correspond to the experimental wavenumber at 1564 cm⁻¹. The scissoring and other vibrational modes have been obtained at lower wavenumbers. The modes obtained at 3537, 3318, and 1645 cm⁻¹ are associated with the NH stretching modes of triazole ring. These vibrations, being purely stretching, possess the PED contributions of 99%, which match with the experimental peaks determined at 3258 and 1639 cm⁻¹. Likewise, the NN stretching vibration of the triazole ring is obtained at 1061 cm⁻¹ with the PED of 53%. Several other vibrational modes including scissoring, twisting, *etc.*, have been obtained at the lower wavenumber than 1000 cm⁻¹.

Group Vibrations

In 4AHT, the phenyl and the triazole rings are bridged by a methylene (-CH₂-) group. Besides, the –OH and -NH₂ groups are substituted in the phenyl and the triazole rings, respectively. The CH₂ stretching modes in 4AHT are obtained at 2996 and 2949 cm⁻¹ having a PED contribution of 99%, which are in accordance with the experimentally observed peak at 2936 cm⁻¹. This is also in agreement with the literature which report that non-aromatic CH stretching vibrations are usually identified in the region lower than the aromatic CH stretching [31]. The CH₂ scissoring modes mixed appreciably with other vibrations are obtained at 1425 cm⁻¹ and 1421 cm⁻¹ having the PED contributions of 50% and 27%, respectively.

The -NH₂ group has the NH stretching vibrations in the lower wavenumber region than those of the NH stretching associated with the ring. The modes obtained at 3390, 3319, and 1645 cm⁻¹ are assigned to the pure anti-symmetric NH₂ stretching of 4AHT, having a PED of 99%, which agrees well with the experimentally measured wavenumbers, 3258 and 1639 cm⁻¹. These modes agree well with the previously reported NH stretching vibrations in the range 3350-3335 cm⁻¹ [37]. Other vibrational modes associated with -NH₂ are calculated in the region beneath

1100 cm^{-1}. The peak obtained at 3674 cm^{-1} corresponds to the OH vibration, which is purely a stretching vibration having 100% PED value and in accordance with the previous studies on several molecules [35 - 37]. The scissoring and torsion modes related to the -OH can be found in the region beneath 500 cm^{-1}.

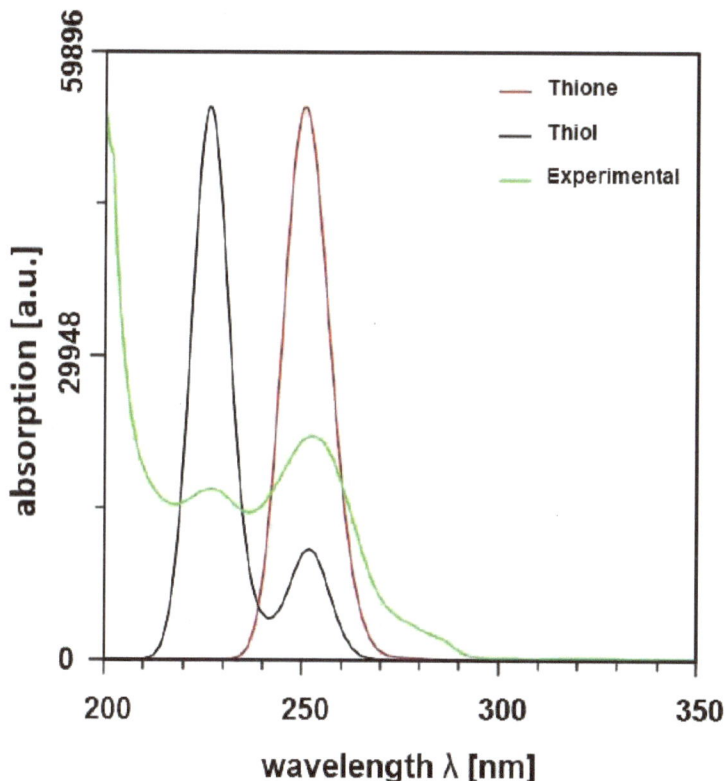

Fig. (6). Experimental UV-visible spectrum of 4AHT and calculated UV-visible spectra of thione and thiol tautomers (reproduced from [19] with the permission of Elsevier).

UV-Visible Spectroscopic Analysis

Fig. (**6**) shows the UV-visible spectrum of 4AHT determined at room temperature using ethanol solvent. It can be seen that the strongest absorption peak is determined at λ_{max} = 252 nm having an absorbance of 1.539 a.u. Two more experimental peaks are measured at 270 and 242 nm. The TD-DFT is employed to simulate the UV-vis spectra with the ethanol solvent, also displayed in Fig. (**6**). The first five excited states of 4AHT of both thione and thiol tautomers have been considered. Table **4** lists the excitation energies, absorption wavelengths, and oscillator strengths obtained in all these excited states. The oscillator strength is a crucial parameter, which measures the absorptivity and the intensity of an electronic transition, *i.e.*, how likely a given electronic transition is permitted (see

Chapter 2). The assignments to these electronic transitions have been made based on the significant role of molecular orbitals and the orbitals having contributions $\leq$ 10% have been omitted.

Table 4. Calculated and scaled wavenumbers, infrared intensity and their assignments. The corresponding FT-IR wavenumbers and intensities are also included for comparison (reproduced from [19] with the permission of Elsevier).

Excitation Energy	Wavelength (nm)		Oscillator Strength	Orbital Transition
(eV)	Calculated	Experimental	(a.u.)	
4.62	268	270	0	HOMO-2→LUMO+1 (56%)
				HOMO-2→ LUMO+2 (23%)
				HOMO-1→ LUMO+1 (12%)
4.92	252	252	0.05	HOMO+1→LUMO (56%)
				HOMO→LUMO (26%)
5.01	248		0.08	HOMO+1→LUMO (22%)
				HOMO→LUMO (70%)
5.06	245	242	0.18	HOMO→ LUMO+1 (90%)
5.14	241	-	0.25	HOMO→ LUMO+2 (87%)

The molecular orbitals participating for electronic transitions in different excited states and corresponding energy eigenvalues are shown in Fig. (7). The first excited state having excitation energy of 4.62 eV and zero oscillator strength includes HOMO-2→LUMO+1 and HOMO-2→LUMO+2 transitions having contributions of 56% and 23%, respectively, corresponding to the experimentally observed peak at 270 nm. The band calculated at 252 nm by TD-DFT method agrees well with the observed band in the UV-visible spectrum. This band results in the second excited state and corresponds to the HOMO+1→LUMO (56%) and HOMO→LUMO (26%). Likewise, the same HOMO→LUMO orbital transition contributes to the third excited state with the contribution of 70%. The frontier orbitals, *i.e.*, HOMO and LUMO are known to play a crucial role in the chemical interaction or reactions with other compounds. It is evident that the HOMO of 4AHT consists of the triazole ring but its LUMO is mainly contributed by the phenyl ring system (see Fig. 7). Consequently, the transition from HOMO to LUMO leads to the transfer of charge transfer from the triazole ring to the phenyl ring system. The energy gap between HOMO and LUMO, E_{gap}, determines the chemical reactivity of molecules. The smaller E_{gap} suggests more chemically reactive nature of molecular systems.

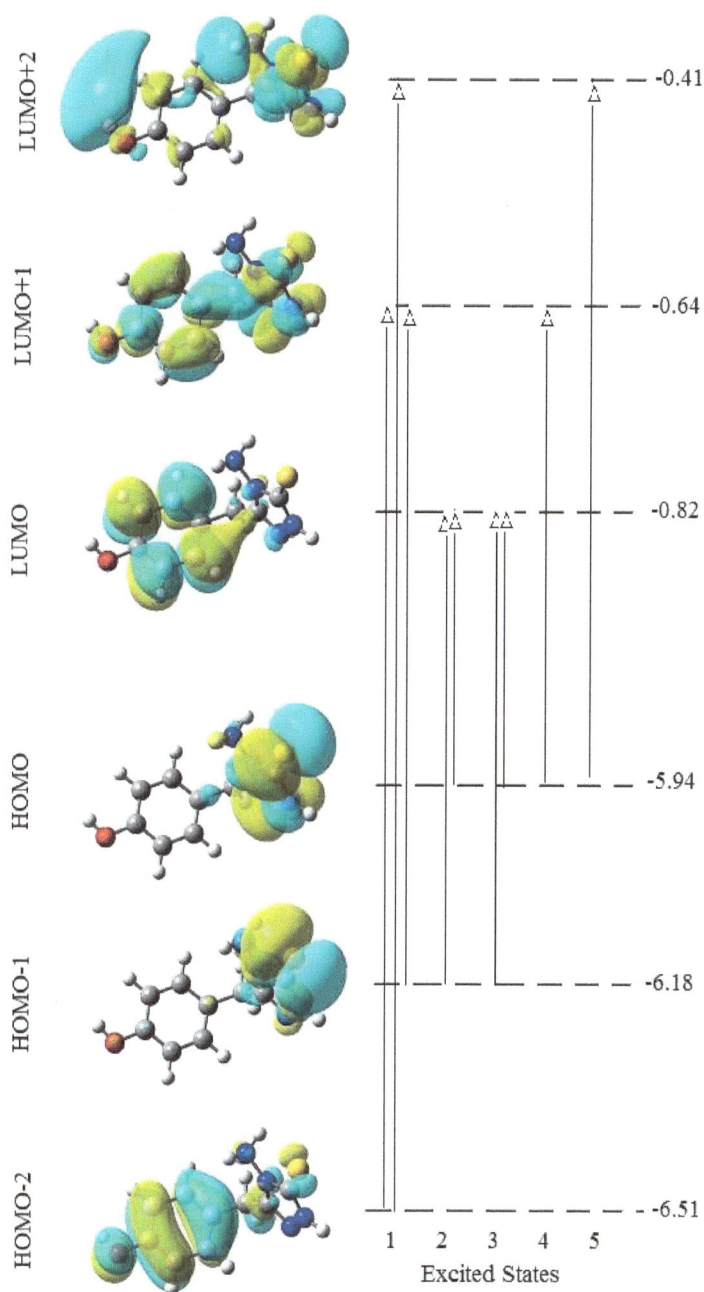

Fig. (7). Electronic transitions in various excited states of 4AHT calculated at TD-DFT//B3LYP/6-31+G(d,p) method (reproduced from [19] with the permission of Elsevier).

The E_{gap} of 4AHT is calculated to be 5.16 eV, which provides the strength of the charge-transfer interactions. The higher excited states of 4AHT lead to the orbital transitions HOMO→LUMO+1 and HOMO→LUMO+2 with the contributions of 90% and 87%, respectively, which is in accordance with the experimental peak observed at 242 nm. The TD-DFT calculations have been also performed on the thiol tautomer. The excited states of thiol form are obtainted at 252, 240, 230, 227, 225, and 221 nm. It is interesting to see that there also exists the electronic absorption peak at 252 nm in thiol tautomer, corresponding to the experimentally observed peak in the UV-vis spectrum. Nevertheless, other absorption wavelengths of thiol form are very close to experimentally observed bands. This suggests the preference of the thiol over thione form, which we call the NH-SH tautomerism.

NMR Spectroscopic Analysis

Fig. (8) displays the observed ^{1}H-NMR spectrum of 4AHT, which was recorded using the DMSO solvent. To explain and discuss this NMR spectrum, we have obtained the chemical shifts of the H nuclei, *i.e.*, proton as mentioned in previous chapters. Just for reference, the isotropic magnetic shielding (IMS) of the optimized structure of tetramethylsilane (TMS) using the present computational scheme is calculated to be 31.6 ppm. Table **5** lists the ^{1}H-NMR experimental chemical shifts of 4AHT along with their calculated values for thione and thiol forms and assigments. The singlet peak of hydrogen linked to nitrogen of the triazole ring (Tr) in thione form is obtained at 9.2 ppm, which matches well with the experimental value (9.3 ppm). In contrast to this, the respective peak of hydrogen attached to sulphur in thiol form is found at 4.8 ppm. It implies that the thione form of 4AHT should dominate in the DMSO solvent. The chemical shifts of hydrogen associated with the aromatic ring (Ar) are found in the range 6.9-7.6 ppm in both tautomers of 4AHT. These chemical shifts are in accordance with the experimental values (6.7-7.1 ppm) and the values reported in literature [38, 39]. The chemical shifts of the hydrogen associated with the -NH$_2$ and -CH$_2$ groups are reduced due to stronger shielding [40]. This is due to the fact that the hydrogen linked to or in the vicinity of an electron-donating group or atom strengthens the shielding and causes to lower the resonant frequency. The singlet peaks at 3.6-4.4 ppm (thione) and 3.9-4.3 ppm (thiol) correspond to the hydrogen of the -NH$_2$ group linked at triazole ring, consistent with the experimental shift of 3.8 ppm. The hydrogen atoms associated with the -CH$_2$- group have singlet as well as doublet peaks at 3.9 ppm, corresponding to the experimental value of 3.3 ppm. The chemical shift of hydrogen associated with the primary -OH group is obtained at 4.7 ppm, which is experimentally determined at 5.5 ppm because of the intermolecular hydrogen bonding in the solvent.

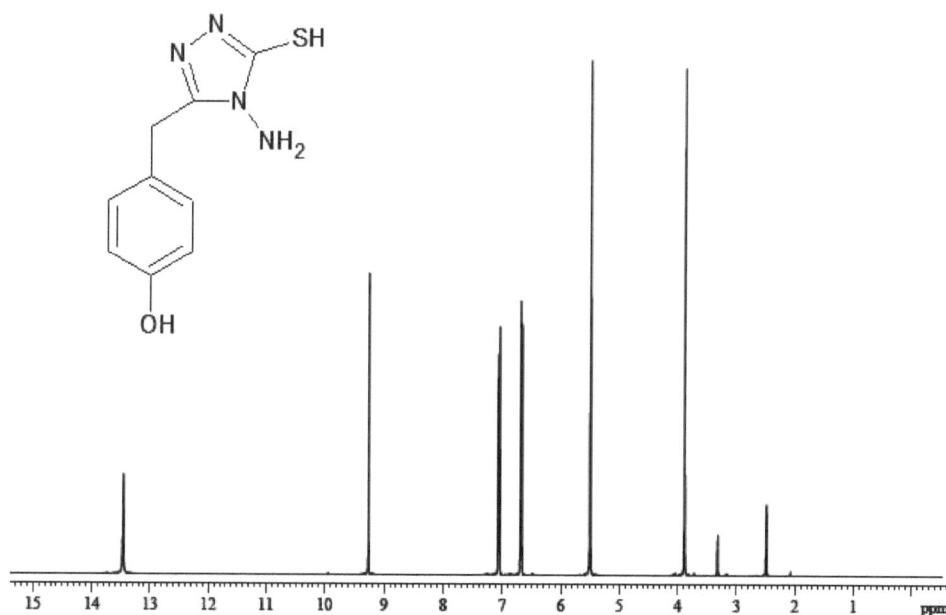

Fig. (8). ^{1}H-NMR spectrum of 4AHT recorded in DMSO solvent (reproduced from [19] with the permission of Elsevier).

Table 5. ^{1}H-NMR chemical shifts (δ) of 4AHT and their assignments (reproduced from [19] with the permission of Elsevier).

Atom	δ (ppm)			Assignment
	Thione	**Thiol**	**Experiment**	
2H	9.2	4.8	9.3	[s, H(Tr)]
9H	7.6	7.5	7.1	[s, H(Ar)]
10H	6.9	6.9	6.7	[s, H(Ar)]
11H	7.1	7	7.1	[s, H(Ar)]
12H	7.4	7.4	7.1	[s, H(Ar)]
19H	3.6	3.9	3.8	[s, H(-NH$_2$)]
20H	4.4	4.3	--	[s, H(-NH$_2$)]
22H	3.9	4.1	3.3	[d, H(-CH$_{2-}$)]
23H	3.9	--	3.3	[s, H(-CH$_{2-}$)]
25H	4.7	--	5.5	[s, H(-OH)]

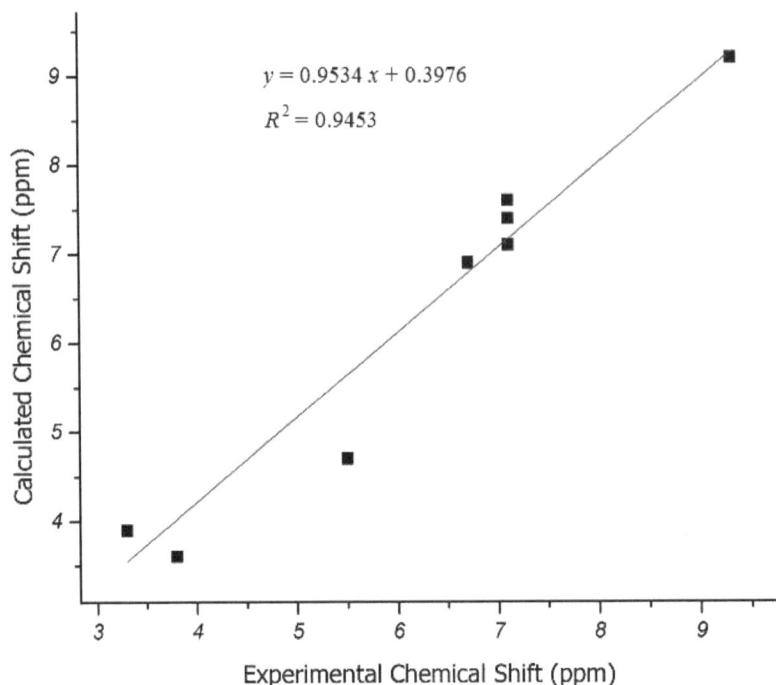

Fig. (9). Correlation graph between calculated and experimental chemical shifts (reproduced from [19] with the permission of Elsevier).

The observed and calculated ^{1}H-NMR chemical shifts have been compared by a statistical correlation between the two as shown in Fig. (**9**). There exists a linear correlation with an equation, $y = 0.9534 \ x + 0.3976$, where y and x are the calculated and the experimentally observed ^{1}H-NMR chemical shifts, respectively. The correlation coefficient of $R^2 = 0.9453$ suggests that the calculated ^{1}H-NMR chemical shifts agree well with the corresponding experimental values. Some aberration appeared between calculated and experimental peaks may be presumably due to the presence of intermolecular interactions, which have been ignored because of the restriction of our calculations on an isolated molecule.

MESP and Electronic Parameters

The molecular electrostatic potential (MESP) is very important in locating the potential sites for electrophilic and nucleophilic substitutions [41]. The MESP also plays a key role in understanding the phenomena of the molecular recognition, *e.g.*, in drug-receptor interactions, due to the fact two species face each other through their potentials [42]. Fig. (**10**) plots the MESP surface of 4AHT using a color-grading. For 4AHT, this color grade ranges from -0.06958 a.u. for the darkest red to +0.06958 a.u. for the darkest blue, where red and blue represent the highly electronegative and highly electropositive potentials regions, respectively. The MESP plot of 4AHT clearly reflects that the hydrogen of the -OH group attached to the phenyl ring happens to be the highly electropositive, *i.e.*, it is a favourable site for the electrophilic substitution in the molecule.

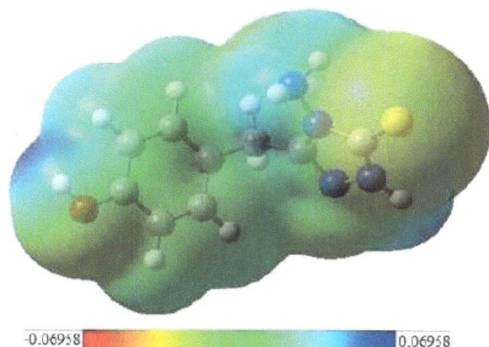

-0.06958   0.06958

Fig. (10). Molecular electrostatic potential (MESP) surface of 4AHT (reproduced from [19] with the permission of Elsevier).

The chemical reactivity of 4AHT has been further discussed by calculating several electronic parameters, namely ionization potentials (*I*), electron affinity (*A*), absolute electronegativity (χ), and chemical hardness (η), *etc*. *I* and *A* values can be obtained by the negative of energy eigenvalues of HOMO and LUMO, respectively (see Chapter 2). Subsequently, the finite-difference approximations [43] are used to calculate χ and η parameters. Table **6** lists these electronic parameters for both thione and thiol forms. It is evident that thione form of 4AHT has a bit smaller *I* but lager *A* as well as η values than its thiol form. It implies the more chemically reactive nature of thione than its thiol tautomer. Parr *et al.* [44] suggested another parameter known as the electrophilic index (ω), also listed in Table **6**. This parameter estimates the stabilization energy of a molecular system due to an extra electronic charge from its neighbourhood. Therefore, ω determines the potency of an electrophile to attain extra charge and the defiance of the molecule to exchange the charge with the neighbourhood. It has been reported that the good electrophiles usually possess higher ω value [45]. However, it can be noticed from Table **6** that ω values of both thione and thione tautomers of 4AHT are almost the same. The dipole moment (μ) of a molecule gives a stamp of its structure and the charge distribution on its constituent atoms. The μ in thione form of 4AHT is found to be 4.04 Debye, which is lower than that of thiol form, *i.e.*, 5.84 Debye. This may suggest that thiol tautomer has more polarized nature as compared to thione form.

Table 6. Electronic parameters of 4AHT in thione and thiol forms (reproduced from [19] with the permission of Elsevier).

Parameter	Thione form	Thiol form
Ionization potential (eV)	5.94	6.28
Electron affinity (eV)	0.82	0.75
Absolute electronegativity (eV)	3.38	3.52
Chemical hardness (eV)	2.56	2.77
Electrophilic index (eV)	2.23	2.24
Dipole moment (Debye)	4.04	5.84

CONCLUSIONS AND REMARKS

To conclude it can be said that this chapter introduced experimental synthesis of 4-Amino-3-(4-hydroxybenzyl)-1H-1,2,4-triazole-5(4H)-thione (4AHT) and its characterization using the FT-IR, ^{1}H-NMR, and UV-visible spectra along with X-ray crystallography. An exhaustive analysis on 4AHT has been performed in its thione and thiol tautomeric forms using density functional theory (DFT). The thermodynamics of reaction supported the route of synthesis. The calculated geometrical parameters show good conformity with the respective experimental values. The FT-IR spectrum of 4AHT has been reported and analyzed by assigning the normal modes of vibration on the basis of potential energy distributions. The UV-visible spectrum of the compound was analyzed and discussed using the TD-DFT approach by calculating the electronic transitions for several excited states. The experimental peak measured in the UV-visible spectrum is perfectly simulated by our calculations. The ^{1}H-NMR spectrum of compound has also been interpreted using the GIAO approach. A good statistical correlation between calculated and experimental chemical shifts has been obtained having a linear correlation coefficient greater than 0.9. The spectroscopic analyses not only indicate the preference of 4AHT for thione form in the solid-state but also suggest the NH-SH tautomerism in the liquid form.

CONSENT FOR PUBLICATION

Not applicable.

CONFLICT OF INTEREST

The authors declare no conflict of interest, financial or otherwise.

ACKNOWLEDGEMENTS

Declared none.

REFERENCES

[1]   Benzon, F.R. Tetrazoles, tetrazines and purines and related ring systems.*Heterocyclic Compounds*; Elderfield, R.C., Ed.; Wiley: New York, **1967**, 8, .

[2]   Temple, C., Jr Triazoles 1,2,4.*Chemistry of the Heterocyclic Compounds*; Montogomery, J.A., Ed.; Wiley: New York, **1981**, p. 37.
       [http://dx.doi.org/10.1002/9780470187104]

[3]   Shivarama Holla, B.; Sooryanarayana Rao, B.; Sarojini, B.K.; Akberali, P.M.; Suchetha Kumari, N. Synthesis and studies on some new fluorine containing triazolothiadiazines as possible antibacterial, antifungal and anticancer agents. *Eur. J. Med. Chem.,* **2006**, *41*(5), 657-663.
       [http://dx.doi.org/10.1016/j.ejmech.2006.02.001] [PMID: 16616396]

[4]   Holla, B. S.; Sarojini, B. K.; Rao, B. S.; Akberali, P. M.; Suchetha Kumari, N.; Shetty, V. >*Synthesis of some halogen-containing 1, 2, 4-triazolo-1, 3, 4-thiadiazines and their antibacterial and anticancer screening studies-Part I, II Farmaco,* **2001**, *56*, 565-570.

[5]   Mullican, M.D.; Wilson, M.W.; Connor, D.T.; Kostlan, C.R.; Schrier, D.J.; Dyer, R.D. Design of 5-(3,5-di-tert-butyl-4-hydroxyphenyl)-1,3,4-thiadiazoles, -1,3,4-oxadiazoles, and -1,2,4-triazoles as orally-active, nonulcerogenic antiinflammatory agents. *J. Med. Chem.,* **1993**, *36*(8), 1090-1099.
       [http://dx.doi.org/10.1021/jm00060a017] [PMID: 8478906]

[6]   Jones, D.H.; Slack, R.; Squires, S.; Wooldridge, K.R.H. Antiviral chemotherapy. I. The activity of pyridine and quinoline derivatives against neurovaccinia in mice. *J. Med. Chem.,* **1965**, *8*, 676-680.
       [http://dx.doi.org/10.1021/jm00329a026]

[7]   Shams el-Dine, S.A.; Hazzaa, A.A.B. Synthesis of compounds with potential fungicidal activity. Part 2. *Pharmazie,* **1974**, *29*(12), 761-763.
       [PMID: 4460001]

[8]   Kane, J.M.; Dudley, M.W.; Sorensen, S.M.; Miller, F.P. 2,4-Dihydro-3H-1,2,4-triazole-3-thiones as potential antidepressant agents. *J. Med. Chem.,* **1988**, *31*(6), 1253-1258.
       [http://dx.doi.org/10.1021/jm00401a031] [PMID: 3373495]

[9]   Asan, A.; Soylu, S.; Kıyak, T.; Yıldırım, F.; Oztas, S.G.; Ancın, N.; Kabasakaloğlu, M. Investigation on some Schiff bases as corrosion inhibitors for mild steel. *Corros. Sci.,* **2006**, *48*, 3933-3944.
       [http://dx.doi.org/10.1016/j.corsci.2006.04.011]

[10]   Yurt, A.; Balaban, A.; Ustun, K.; Bereket, G.; Erk, B. Investigation on some Schiff bases as HCl corrosion inhibitors for carbon steel. *Mater. Chem. Phys.,* **2004**, *85*, 420-426.
       [http://dx.doi.org/10.1016/j.matchemphys.2004.01.033]

[11]   Ramesh, S.; Rajeswari, S. Corrosion inhibition of mild steel in neutral aqueous solution by new triazole derivatives. *Electrochim. Acta,* **2004**, *49*, 811-820.
       [http://dx.doi.org/10.1016/j.electacta.2003.09.035]

[12]   Yurt, A.; Ulutas, S.; Dal, H. Electrochemical and theoretical investigation on the corrosion of aluminium in acidic solution containing some Schiff bases. *Appl. Surf. Sci.,* **2006**, *253*, 919-925.
       [http://dx.doi.org/10.1016/j.apsusc.2006.01.026]

[13]   Agrawal, Y.K.; Talati, J.D.; Shah, M.D.; Desai, M.N.; Shah, N.K. Schiff bases of ethylenediamine as corrosion inhibitors of zinc in sulphuric acid. *Corros. Sci.,* **2004**, *46*, 633-651.
       [http://dx.doi.org/10.1016/S0010-938X(03)00174-4]

[14]   Bentiss, F.; Bouanis, M.; Mernari, B.; Traisnel, M.; Vezin, H.; Lagrenee, M. Understanding the adsorption of 4H-1, 2, 4-triazole derivatives on mild steel surface in molar hydrochloric acid. *Appl.*

Surf. Sci., **2007**, *253*, 3696-3704.
[http://dx.doi.org/10.1016/j.apsusc.2006.08.001]

[15]   Al Ashry, E.S.H.; El Nemr, A.; Esawy, S.A.; Ragab, S. Corrosion inhibitors-Part II: Quantum chemical studies on the corrosion inhibitions of steel in acidic medium by some triazole, oxadiazole and thiadiazole derivatives. *Electrochim. Acta,* **2006**, *51*, 3957-3968.

[16]   Wang, H.L.; Liu, R.B.; Xin, J. Inhibiting effects of some mercapto-triazole derivatives on the corrosion of mild steel in 1.0 M HCl medium. *Corros. Sci.,* **2004**, *46*, 2455-2466.
[http://dx.doi.org/10.1016/j.corsci.2004.01.023]

[17]   Wang, H.L.; Fan, H.B.; Zheng, J.S. Corrosion inhibition of mild steel in hydrochloric acid solution by a mercapto-triazole compound. *Mater. Chem. Phys.,* **2003**, *77*, 655-661.
[http://dx.doi.org/10.1016/S0254-0584(02)00123-2]

[18]   Salgin-Gökşen, U.; Gökhan-Kelekçi, N.; Göktaş, O.; Köysal, Y.; Kiliç, E.; Işik, S.; Aktay, G.; Özalp, M. 1-Acylthiosemicarbazides, 1,2,4-triazole-5(4H)-thiones, 1,3,4-thiadiazoles and hydrazones containing 5-methyl-2-benzoxazolinones: synthesis, analgesic-anti-inflammatory and antimicrobial activities. *Bioorg. Med. Chem.,* **2007**, *15*(17), 5738-5751.
[http://dx.doi.org/10.1016/j.bmc.2007.06.006] [PMID: 17587585]

[19]   Srivastava, A.K.; Kumar, A.; Misra, N.; Manjula, P.S.; Sarojini, B.K.; Narayana, B. Synthesis, spectral (FT-IR, UV-visible, NMR) features, biological activity prediction and theoretical studies of 4-Amin--3-(4-hydroxybenzyl)-1H-1,2,4-triazole-5(4H)-thione and its tautomer. *J. Mol. Struct.,* **2016**, *1107*, 137-144.
[http://dx.doi.org/10.1016/j.molstruc.2015.11.042]

[20]   Sarojini, B.K.; Manjula, P.S.; Kaur, M.; Anderson, B.J.; Jasinski, J.P. 4-Amino-3-(4-hy-drox--benz-yl)-1H-1,2,4-triazole-5(4H)-thione. *Acta Crystallogr. Sect. E Struct. Rep. Online,* **2013**, *70*(Pt 1), o48-o49.
[PMID: 24526991]

[21]   Becke, A.D. Density-functional exchange-energy approximation with correct asymptotic behavior. *Phys. Rev. A Gen. Phys.,* **1988**, *38*(6), 3098-3100.
[http://dx.doi.org/10.1103/PhysRevA.38.3098] [PMID: 9900728]

[22]   Lee, C.; Yang, W.; Parr, R.G. Development of the Colle-Salvetti correlation-energy formula into a functional of the electron density. *Phys. Rev. B Condens. Matter,* **1988**, *37*(2), 785-789.
[http://dx.doi.org/10.1103/PhysRevB.37.785] [PMID: 9944570]

[23]   Barone, V.; Cossi, M.; Tomasi, J. A new definition of cavities for the computation of solvation free energies by the polarizable continuum model. *J. Chem. Phys.,* **1997**, *107*, 3210-3221.
[http://dx.doi.org/10.1063/1.474671]

[24]   Frisch, M.J.; Trucks, G.W.; Schlegel, H.B. *Gaussian 09, Revision B.01*; Gaussian Inc: Wallingford, CT, **2010**.

[25]   Romani, D.; Brandan, S.A. Structural and spectroscopic studies of two 1, 3-benzothiazole tautomers with potential antimicrobial activity in different media. Prediction of their reactivities. *Comput. Theor. Chem.,* **2015**, *1061*, 89-99.
[http://dx.doi.org/10.1016/j.comptc.2015.03.018]

[26]   Alecu, I.M.; Zheng, J.; Zhao, Y.; Truhlar, D.G. Computational thermochemistry: scale factor databases and scale factors for vibrational frequencies obtained from electronic model chemistries. *J. Chem. Theory Comput.,* **2010**, *6*(9), 2872-2887.
[http://dx.doi.org/10.1021/ct100326h] [PMID: 26616087]

[27]   Alcolea Palafox, M. Scaling factors for the prediction of vibrational spectra. I. Benzene molecule. *Int. J. Quantum Chem.,* **2000**, *77*, 661-684.
[http://dx.doi.org/10.1002/(SICI)1097-461X(2000)77:3<661::AID-QUA7>3.0.CO;2-J]

[28]   Jamroz, M. H. *Vibrational energy distribution analysis VEDA 4,* **2004**.

[29]   Jamróz, M.H. Vibrational energy distribution analysis (VEDA): scopes and limitations. *Spectrochim. Acta A Mol. Biomol. Spectrosc.,* **2013**, *114*, 220-230.
[http://dx.doi.org/10.1016/j.saa.2013.05.096] [PMID: 23778167]

[30]   Dennington, R.; Keith, T.; Millam, J. *GaussView Version 5.0*; Semichem Inc: KS, **2005**.

[31]   George, S. *Infrared and Raman characteristic group frequencies – Tables and Charts,* 3rd ed; Wiley: Chichester, **2001**.

[32]   Silverstein, M.; Basseler, G.G.; Morrill, C. *Spectrometric identification of Organic Compounds*; Wiley: New York, **1991**.

[33]   Arivazhagan, M.; Senthil kumar, J. Vibrational analysis of 4-amino pyrazolo (3,4-d) pyrimidine A joint FTIR, Laser Raman and scaled quantum mechanical studies. *Spectrochim. Acta A Mol. Biomol. Spectrosc.,* **2011**, *82*(1), 228-234.
[http://dx.doi.org/10.1016/j.saa.2011.07.040] [PMID: 21824808]

[34]   Srivastava, A.K.; Narayana, B.; Sarojini, B.K.; Misra, N. Vibrational, structural and hydrogen bonding analysis of N-[(E)-4-hydroxybenzylidene]-2-(naphthalen-2-yloxy) acetohydrazide: Combined density functional and atoms-in-molecule based theoretical studies. *Indian J. Phys.,* **2014**, *88*, 547-556.
[http://dx.doi.org/10.1007/s12648-014-0449-y]

[35]   Srivastava, A.K.; Baboo, V.; Narayana, B.; Sarojini, B.K.; Misra, N. Comparative DFT study on reactivity, acidity and vibrational spectra of halogen substituted phenylacetic acids. *Indian J. Pure Appl. Phy.,* **2014**, *52*, 507-519.

[36]   Dwivedi, A.; Srivastava, A.K.; Bajpai, A. Vibrational spectra, HOMO, LUMO, MESP surfaces and reactivity descriptors of amylamine and its isomers: A DFT study. *Spectrochim. Acta A Mol. Biomol. Spectrosc.,* **2015**, *149*, 343-351.
[http://dx.doi.org/10.1016/j.saa.2015.04.042] [PMID: 25965519]

[37]   Kumar, A.; Srivastava, A.K.; Gangwar, S.; Misra, N.; Mondal, A.; Brahmachari, G. Combined experimental (FT-IR, UV–visible spectra, NMR) and theoretical studies on the molecular structure, vibrational spectra, HOMO, LUMO, MESP surfaces, reactivity descriptor and molecular docking of Phomarin. *J. Mol. Struct.,* **2015**, *1096*, 94-101.
[http://dx.doi.org/10.1016/j.molstruc.2015.04.031]

[38]   Smith, W.B.; Ihrigz, A.M.; Roark, J.L. Substituent effects on aromatic proton chemical shifts. VII. Further examples drawn from disubstituted benzenes. *J. Phys. Chem.,* **1970**, *74*, 812-821.
[http://dx.doi.org/10.1021/j100699a024]

[39]   Wang, B.; Fleischer, U.; Hinton, J.F.; Pulay, P. Accurate prediction of proton chemical shifts. I. Substituted aromatic hydrocarbons. *J. Comput. Chem.,* **2001**, *22*, 1887-1895.
[http://dx.doi.org/10.1002/jcc.1139]

[40]   Cotton, F.A.; Wilkinson, C.W. *Advanced Inorganic Chemistry,* 3rd ed; Interscience publisher: New York, **1972**.

[41]   Luqul, F.J.; Lopez, J.M.; Orozco, M. Electrostatic interactions of a solute with a continuum. A direct utilization of ab initio molecular potentials for the prevision of solvent effects. *Theor. Chem. Acc.,* **2000**, *103*, 343-345.
[http://dx.doi.org/10.1007/s002149900013]

[42]   Scrocco, E.; Tomasi, J. The electrostatic molecular potential as a tool for the interpretation of molecular properties. *Top. Curr. Chem.,* **1973**, *7*, 95-170.
[http://dx.doi.org/10.1007/3-540-06399-4_6]

[43]   Parr, R.G.; Yang, W. *W. Density Functional Theory of Atoms and Molecules*; Oxford University Press and Clarendon Press: New York, Oxford, **1989**.

[44]   Parr, R.G.; Szentpaly, L.V.; Liu, S. Electrophilicity index. *J. Am. Chem. Soc.,* **1999**, *121*, 1922-1924.
[http://dx.doi.org/10.1021/ja983494x]

[45]   Parthasarathi, R.; Subramanian, V.; Roy, D.R.; Chattaraj, P.K. Electrophilicity index as a possible descriptor of biological activity. *Bioorg. Med. Chem.,* **2004**, *12*(21), 5533-5543.
[http://dx.doi.org/10.1016/j.bmc.2004.08.013] [PMID: 15465330]

CHAPTER 7

Inter- and Intra-Molecular Interactions by Quantum Theory of Atoms in Molecule

Abstract: In this exclusive chapter, we present a brief overview of the quantum theory of atoms in molecule (QTAIM) proposed by R. F. W. Bader. This theory is based on the topological analysis of the electron density and related parameters. One of the strengths of this theory is the accurate prediction, characterization, and quantification of various interactions, including H-bond and van der Waals interactions. Herein, we discuss the important aspects of QTAIM regarding the intra- and intermolecular interactions in biologically active molecules discussed in the preceding chapters. Within the framework of the QTAIM, I mention the criteria of H-bonds and characterize the various H-bonds on the basis of topological parameters, and continue to quantify the strength of the H-bond. I will also describe a very user-friendly software *AIMAll* to perform the QTAIM analysis and explain the obtained results. In order to provide the contents digestive, I will include examples of molecules from Chapters 3, 4, and 5. I believe that this chapter will guide the readers interested in various kinds of interactions in biologically active molecules.

Keywords: AIM, AIMAll, BCP, Bond strength, BSSE, Characterization, Complex, Counterpoise, Dimer, Electron density, H-bond, Intermolecular interaction, Intramolecular interaction, Laplacian, Molecular graph, QTAIM, RCP, Reactivity, Stabilization, Supermolecular approach, Topology.

INTRODUCTION

H-bonds play a key role in determining the shapes, properties, and functions of biomolecules and biologically active systems [1]. Despite the dominant role of H-bonding in nature, accurate data on the respective stabilization or interaction energies are quite rare. The situation with extended H-bonded complexes is, despite enormous progress in various experimental techniques, even less satisfactory and accurate data on stabilization energies of these complexes are almost unavailable. Reliable and consistent information on the stability of various types of inter-molecular H-bonded complexes, from the very weak to the strongest, comes from the high-level correlated quantum chemical wavefunction-based method or density functional theory (DFT) calculations, and these methods, thus, represent one of the most promising sources of relevant data. On the cont-

Ambrish Kumar Srivastava and Neeraj Misra

rary, the intra-molecular H-bonds can not be directly quantified from a DFT calculation and this requires an alternative approach.

QUANTUM THEORY OF ATOMS IN MOLECULE (QTAIM)

The QTAIM [2] describes the interactions between atomic basins. It exploits some topological parameters *viz.* electron density (ρ) and its Laplacian $(\nabla^2 \rho)$, kinetic energy density (G), potential energy density (V), and total electron energy density (H) at the bond critical point (BCP) of interaction atoms or fragments. The theory efficiently characterizes and quantifies various types of H-bonded and other interactions.

Existence of H-Bond

In QTAIM, the existence of H-bond follows Koch and Popelier criterion [3], which requires:

1. The existence of bond critical point (BCP) for the 'proton (H)···acceptor (A)' contact.
2. The value of electron density should lie in the range of 0.002–0.040 a.u.
3. The corresponding Laplacian ($\nabla^2 \rho$) should be within the range of 0.024–0.139 a.u.

Characterization of H-Bond

The three types of H-bond are characterized on the basis of topological parameters. According to Rozas *et al.* [4], the characterization demands at BCP:

1. $\nabla^2 \rho < 0$ and $H < 0$ for strong H-bond of covalent nature.
2. $\nabla^2 \rho > 0$ and $H < 0$ for medium H-bond of partially covalent nature.
3. $\nabla^2 \rho > 0$ and $H > 0$ for weak H-bond of electrostatic character.

Strength of H-Bond

According to Espinosa *et al.* [5], the interaction energy of A···B contact is defined as,

$$\Delta E = -\tfrac{1}{2}V \tag{1}$$

at BCP.

The *AIMAll* Program

There are various programs available to perform the QTAIM analysis, such as *AIMPAC* [6], *AIM2000* [7], *AIMAll* [8, 9], *etc.* In this chapter, we use the *AIMAll* program in order to explore intra-molecular and inter-molecular H-bond interactions such as hydrogen bonds (H-bonds). *AIMAll* is a user-friendly and efficient program. The graphical user interface of *AIMAll* is displayed in Fig. (**1**). It uses the formatted checkpoint (.fch or .fchk) file generated by the *Gaussian* program (see Chapter 2) as an input and automatically performs QTAIM analysis for the system under study. The basis set on reliability and stability in the values of QTAIM parameters have been studied and found that they are almost independent of the basis set in the case of used functional B3LYP in DFT [10].

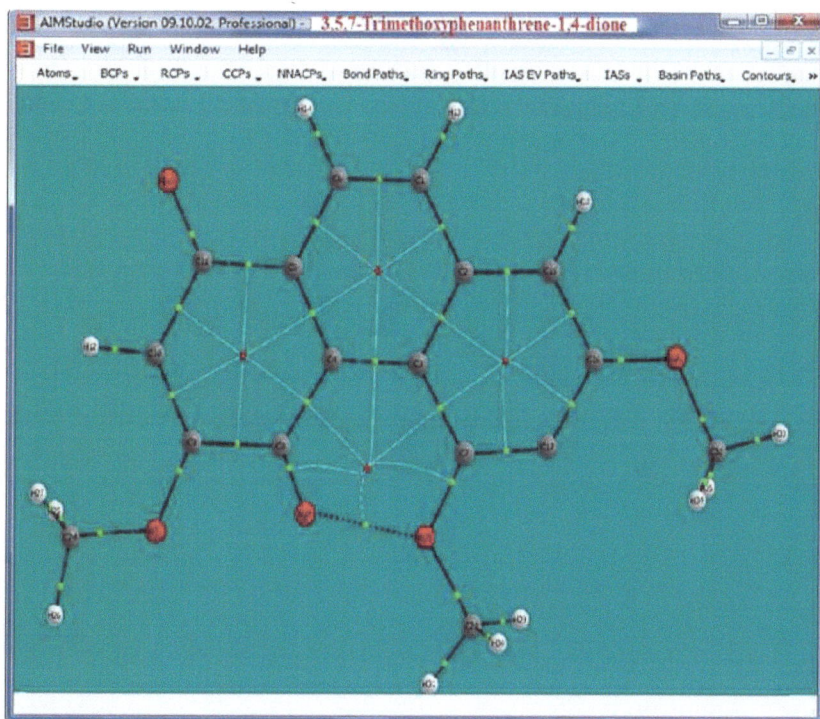

Fig. (1). The graphical user interface of *AIMAll* program displaying the molecular graph of 3,5,7-Trimethoxyphenanthrene-1,4-dione (TPD). The intra-molecular interaction is shown by a dotted line. Green and red points represent BCPs and RCPs, respectively.

The QTAIM calculations on TPD [11] reveal an intra-molecular O20$\cdots$O22 interaction as depicted in the molecular graph shown in Fig. (**1**). The bond-distance O20$\cdots$O22 calculated at B3LYP/6-31+G** method is 2.643 Å. The values of topological parameters at BCP of O22$\cdots$O21 are as follows:

Electron density (ρ) = 0.016446 a.u.

Laplacian ($\nabla^2\rho$) = 0.059157 a.u.

Potential energy density (V) = - 0.013225 a.u.

Kinetic energy density (G) = 0.0140073 a.u.

Total energy density ($H = V + G$) = 0.000782 a.u.

Since $\nabla^2\rho > 0$ and H > 0, the interaction O20···O22 in TPD should be characterized as weak and mainly electrostatic in nature. The interaction energy (ΔE) of O20···O22, by eq. (1) is

ΔE = 4.15 kcal/mol (1 a.u. = 627.15 kcal/mol).

INTRA-MOLECULAR INTERACTIONS

Synthetic Compounds

We perform QTAIM analysis on 26DTA, 24DTA, and 34DTA as discussed in Chapter 3. Table **1** lists the results of the QTAIM analysis. The molecular graphs showing the interaction between O and S in these molecules are displayed in Fig. (**2**). The analyses characterize an ionic interaction between O of the amide fragment and S of the thiazole ring. The partial charges on O and S also suggest the interaction between the two essentially ionic (see Chapter 3). The energy between two interacting atoms can be estimated by eq. (1), which is 4.2 kcal/mol for each molecule. This weak interaction provides very little stabilization in these systems.

Table 1. QTAIM calculations performed on the synthetic compounds discussed in Chapter 3. All parameters are in a.u.

Molecule	Interaction	Length (Å)	ρ	$\nabla^2\rho$	V	G	H	ΔE (kcal/mol)
26DTA	S19···O12	2.650	0.0178	0.0578	-0.0134	0.0139	0.0005	4.2
24DTA	S23···O16	2.303	0.0179	0.0579	-0.0134	0.0139	0.0005	4.2
34DTA	S19···O12	2.757	0.0179	0.0579	-0.0134	0.0139	0.0005	4.2

Natural Products

The existence of C–H···O type H-bonds has been a matter of great controversy for the last many decades [12, 13] in spite of the fact that they play a dominant role in crystal engineering [14] due to their influence on packing motifs.

Furthermore, their biological impact can never be ignored due to their occurrence in carbohydrates [15] and nucleosides [16]. Moreover, CH donors may participate in the coordination of water molecules with the same functionality as OH and NH [17]. We noticed the presence of several C−H···O interactions in the natural products discussed in Chapter 5. The results of QTAIM analyses on triclisine, rufescine, and imerubrine are listed in Table **2** and their molecular graphs are displayed in Fig. (**3**), which shows the intra-molecular interactions by dotted lines.

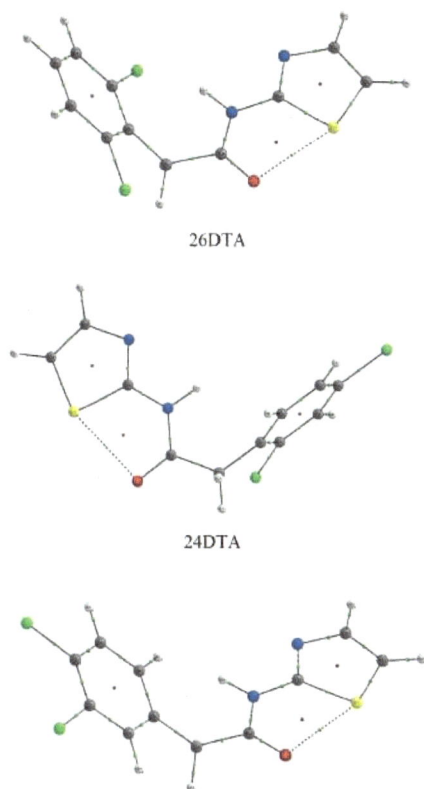

26DTA

24DTA

Fig. (2). Molecular graph of synthetic compounds showing the interaction between O (red sphere) and S (yellow sphere) atoms. Green points represent BCP and red points represent RCP.

There exist two intra-molecular H-bonds in triclisine but three H-bonds in rufescine and imerubrine. The ΔE of H-bonds in triclisine, O28···H22, and O23···H31 are 1.3 and 2.9 kcal/mol, respectively. Similarly, rufescine possesses O26···H20, O21···H29, and O21···H33 with ΔE values of 1.1, 2.7, and 2.8 kcal/mol, respectively. The additional −OCH$_3$ group attached to rufescine leads O21 to act as an acceptor in two H-bonds *viz.* O21···H29 and O21···H33. The total interaction energies are found to be 4.12 kcal/mol in triclisine and 6.7 kcal/mol in rufescine. Further, the ΔE values of O20···H32, O20···H28, and

O25···H19 interactions in imerubrine are calculated to be 2.9, 2.8, and 1.4 kcal/mol, respectively. Evidently, all these interactions are too weak to provide any significant stabilization. However, these interactions are relatively larger in strength as compared to those in rufescine. Thus, the enhanced H-bonding in imerubrine may suggest that it is chemically more reactive as compared to rufescine (see Chapter 5).

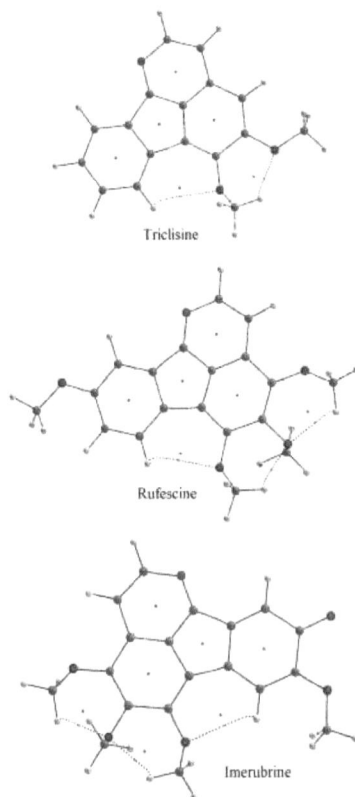

Fig. (3). Molecular graphs showing hydrogen bonding (dotted lines) in some natural products. Red points correspond to ring critical points and green represent bond critical points (BCPs).

Table 2. QTAIM analysis on intra-molecular C−H···O interactions in some natural products discussed in Chapter 5.

Molecule	Interaction	Length (Å)	ρ	∇^2_ρ	V	G	H	ΔE (kcal/mol)
Triclisine	O28···H22	2.760	0.0063	0.0232	-0.0040	0.0049	0.0009	1.3
	O23···H31	2.353	0.0134	0.0511	-0.0091	0.0109	0.0018	2.9

(Table 2) cont.....

Rufescine	O26···H20	2.826	0.0055	0.0207	-0.0035	0.0043	0.0008	1.1
	O21···H29	2.398	0.0129	0.0465	-0.0088	0.0102	0.0014	2.7
	O21···H33	2.374	0.0134	0.0482	-0.0090	0.0105	0.0015	2.8
Imerubrine	O20···H32	2.352	0.0139	0.0504	-0.0094	0.0110	0.0016	2.9
	O20···H28	2.374	0.0134	0.0485	-0.0090	0.0105	0.0015	2.8
	O25···H19	2.710	0.0067	0.0243	-0.0043	0.0052	0.0009	1.4

INTER-MOLECULAR INTERACTIONS

DFT *versus* QTAIM

So far, we have discussed the intra-molecular interactions, which cannot be directly quantified by the DFT method. However, DFT calculations offer a direct estimation of intermolecular interaction energy as energy lost by dimerization or complexation of molecules [18]. This is what is known as the super-molecular approach. The energy of interaction between two molecules A and B in dimer or complex AB can be calculated as,

$$\Delta E = E(AB) - E(A) - E(B) \tag{2}$$

where $E(.)$ denotes the total electronic energies of respective species. ΔE must be corrected to account for basis set superposition errors (BSSE). Boys and Bernadi have suggested a counterpoise (CP) method [19] that ensures simultaneous optimization of geometries of dimeric complexes along with their components and hence, overcome the superposition. The structures (graphs) of molecular systems considered are shown in Fig. (**4**). The interaction energies for H-bonded dimers and complexes calculated by DFT using the CP technique are listed in Table **3**. For comparison, the ΔE values obtained by eq. (1) are also included.

Table 3. Inter-molecular interaction energy for prototype dimers and complexes.

Molecular System	Inter-molecular Interaction Energy (kcal/mol)		
	DFT	QTAIM	Experiment
$(H_2O)_2$	5.08	5.02	5.1
$(HF)_2$	4.64	5.89	5.0-7.0
$(NH_3)_2$	2.88	2.32	4.5
$(H_2O)(HF)$	9.03	10.98	--
$(HF)(NH_3)$	13.24	15.37	--
$(NH_3)(H_2O)$	6.52	5.71	--

We found that the QTAIM calculated values are relatively more reliable, in general [20]. However, in the case of the N–H··N bond, the calculated energies show a fair departure from the experimental values in both approaches, and hence, the corresponding interaction energy values should be handled with care. This study is supposed to assist the theoretical calculations on inter-molecular H-bond interaction energies in large molecular systems where H-bond affects the structure, properties, and functions, remarkably.

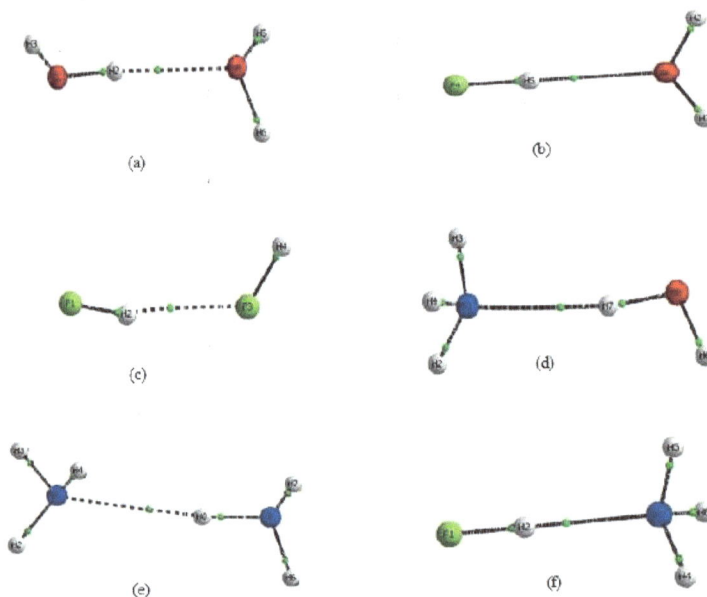

(a)   (b)

(c)   (d)

(e)   (f)

Fig. (4). Molecular structures (graphs) of hydrogen bonded dimers and complexes of H_2O, HF and NH_3.

Inter-Molecular Interactions in CHDP Dimer

We analyze the intermolecular interaction in CHDP dimer as discussed in Chapter 4. The molecular graph of CHDP dimer is shown in Fig. (5) and corresponding parameters are listed in Table 4. QTAIM analysis performed on CHDP dimers reveals two intermolecular interactions *viz.* O34···H13 and O11···H36 in "a" and three inter-molecular interactions *viz.* O32···H13, O11···H26, and O11···H28 in "b." In dimeric form "a," all hydrogen-bonded interactions possess moderate strengths and partially covalent nature, as $\nabla^2\rho) > 0$ and $H) < 0$. On the contrary, in dimeric form "b," all interactions can be characterized as weak and mainly electrostatic due to the $\nabla^2\rho > 0$ but $H > 0$. The calculated ΔE values (Table 4) suggest that dimer "a" is relatively more stabilized due to hydrogen bonding interactions. This is consistent with the DFT calculations and NBO analysis performed on the CHDP dimer, as discussed in Chapter 4.

Table 4. QTAIM parameters calculated for inter-molecular interactions in CHDP dimer discussed in Chapter 4. All parameters are in a.u.

Dimer	A⋯B	Length (Å)	ρ	∇_ρ^2	V	G	H	ΔE (kcal/mol)
a	O34⋯H13	1.645	0.0498	0.1397	-0.0381	0.0365	-0.0016	12.0
	O11⋯H36	1.653	0.0488	0.1379	-0.0371	0.0358	-0.0013	11.6
b	O32⋯H13	1.765	0.0355	0.1099	-0.0255	0.0265	0.0010	8.0
	O11⋯H26	2.837	0.0045	0.0179	-0.0027	0.0036	0.0009	0.9
	O11⋯H28	2.518	0.0091	0.0305	-0.0059	0.0068	0.0009	1.9

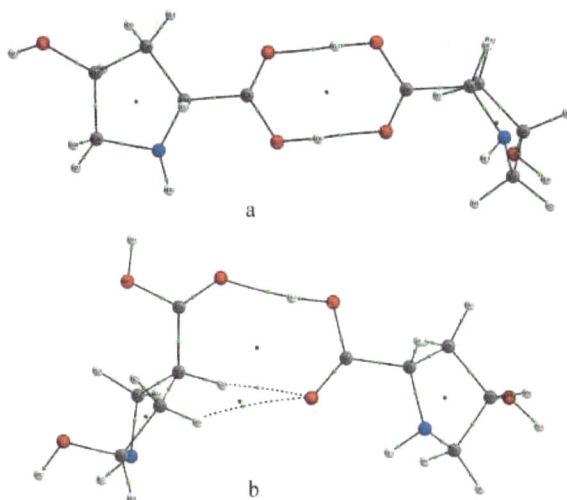

Fig. (5). Molecular graph of CHDP dimer as revealed by the QTAIM analysis.

CONCLUDING REMARKS

In summary, the QTAIM method is capable not only to identify and characterize but also to quantify various types of inter- and intra-molecular interactions. In the framework of QTAIM, an interaction between two atoms is identified by the presence of a BCP between them. The topological parameters at the BCP such as electron density, its Laplacian, kinetic energy density, potential energy density, *etc.* can be used to characterize these interactions. In the case of intramolecular interactions, QTAIM is the only method to deal with them. For intermolecular interactions, QTAIM provides more accurate stabilization energy values than the conventional supermolecular approach of DFT. Although this theory is relatively new, its capabilities are well tested, recognized, and established. There are several software programs based on this theory, *AIMAll*, one among them has been also explored using the molecular systems discussed in this book. Nevertheless, QTAIM is not only limited to H-bonds and other interactions; it touches almost every aspect of the study on molecular systems. An account of this theory can be found from the founder of this theory [2].

CONSENT FOR PUBLICATION

Not applicable.

CONFLICT OF INTEREST

The authors declare no conflict of interest, financial or otherwise.

ACKNOWLEDGEMENTS

Declared none.

REFERENCES

[1]   Jeffrey, G.; Jeffrey, G. A.; Saenger, W.. *Hydrogen Bonding in Biological Structures*; Springer-Verlag: Berlin, Heidelberg, **1991**.
[http://dx.doi.org/10.1007/978-3-642-85135-3]

[2]   Bader, R.F.W. *Atoms in Molecules: A Quantum Theory*; Oxford, New York, **1990**.

[3]   Koch, U.; Popelier, P.L.A. Characterization of C-H-O hydrogen bonds on the basis of the charge density. *J. Phys. Chem.,* **1995**, *99*, 9747-9754.
[http://dx.doi.org/10.1021/j100024a016]

[4]   Rozas, I.; Alkorta, I.; Elguero, J. Behaviour of ylides containing N, O and C atoms as hydrogen bond acceptors. *J. Am. Chem. Soc.,* **2000**, *122*, 11154-11161.
[http://dx.doi.org/10.1021/ja0017864]

[5]   Espinosa, E.; Molins, E.; Lecomte, C. Hydrogen bond strengths revealed by topological analyses of experimentally observed electron densities. *Chem. Phys. Lett.,* **1998**, *285*, 170-173.
[http://dx.doi.org/10.1016/S0009-2614(98)00036-0]

[6]   https://www.chemistry.mcmaster.ca/aimpac/imagemap/imagemap.htm

[7]   http://www.aim2000.de/

[8]   Keith, T.A. *TK Gristmill Software, AIMAll, Version 12.09.23*; Overland Park KS, USA, **2012**.

[9]   http://aim.tkgristmill.com/

[10]   Jabłoński, M.; Palusiak, M. Basis set and method dependence in atoms in molecules calculations. *J. Phys. Chem. A,* **2010**, *114*(5), 2240-2244.
[http://dx.doi.org/10.1021/jp911047s] [PMID: 20085256]

[11]   Brahmachari, G.; Das, S. Biswas (Sinha), M.; Kumar, A.; Srivastava, A. K.; Misra, N. 3,5,7-Trimethoxyphenanthrene-1,4-dione: a new biologically relevant natural phenanthrenequinone derivative from Dioscorea prazeri and studies on its single X-ray crystallographic behavior, molecular docking and other physico-chemical properties. *RSC Advances,* **2016**, *6*, 7317-7329.
[http://dx.doi.org/10.1039/C5RA21490D]

[12]   Taylor, R.; Kennard, O. Crystallographic evidence for the existence of C-H.O, C-H.N and C-H.Cl hydrogen bonds. *J. Am. Chem. Soc.,* **1982**, *104*, 5063-5070.
[http://dx.doi.org/10.1021/ja00383a012]

[13]   Desiraju, G.R. The C-H.O hydrogen bond in crystals: what is it? *Acc. Chem. Res.,* **1991**, *24*, 290-296.
[http://dx.doi.org/10.1021/ar00010a002]

[14]   Aakeroy, C.B.; Sneddon, K.R. The hydrogen bond and crystal engineering. *Chem. Soc. Rev.,* **1993**, *22*, 397-407.
[http://dx.doi.org/10.1039/CS9932200397]

[15]   Steiner, T.; Saenger, W. Geometry of carbon-hydrogen.oxygen hydrogen bonds in carbohydrate crystal structures. Analysis of neutron diffraction data. *J. Am. Chem. Soc.,* **1992**, *114*, 10146-10154.
[http://dx.doi.org/10.1021/ja00052a009]

[16]   Saenger, W. *Principles of Nucleic Acid Structure*; Springer-Verlag: Berlin, Heidelberg, **1984**.
[http://dx.doi.org/10.1007/978-1-4612-5190-3]

[17]   Steiner, T.; Saenger, W. Role of CH.O hydrogen bonds in the coordination of water molecules. Analysis of neutron diffraction data. *J. Am. Chem. Soc.,* **1993**, *115*, 4540-4547.
[http://dx.doi.org/10.1021/ja00064a016]

[18]   Gordon, M.S.; Jensen, J.H. Understanding the hydrogen bond using quantum chemistry. *Acc. Chem. Res.,* **1996**, *29*, 536-543.
[http://dx.doi.org/10.1021/ar9600594]

[19]   Boys, S.F.; Bernardi, F. The calculation of small molecular interactions by the difference of separate total energies. Some procedures with reduced errors. *Mol. Phys.,* **1970**, *19*, 553-566.
[http://dx.doi.org/10.1080/00268977000101561]

[20]   Srivastava, A.K.; Misra, N. Calculating interaction energies of hydrogen bonded dimers and complexes of HF, H2O and NH3: Super-molecular versus AIM Approach. *J. Comput. Methods Mol. Des.,* **2014**, *4*, 19-23.

APPENDIX

Gaussian **Output File of Methane (CH$_4$)**

Entering Link 1 = C:\G09W\l1.exe PID= 3188.

Copyright (c) 1988,1990,1992,1993,1995,1998,2003,2009,2010, Gaussian, Inc. All Rights Reserved.

This is part of the Gaussian(R) 09 program. It is based on

the Gaussian(R) 03 system (copyright 2003, Gaussian, Inc.),

the Gaussian(R) 98 system (copyright 1998, Gaussian, Inc.),

the Gaussian(R) 94 system (copyright 1995, Gaussian, Inc.),

the Gaussian 92(TM) system (copyright 1992, Gaussian, Inc.),

the Gaussian 90(TM) system (copyright 1990, Gaussian, Inc.),

the Gaussian 88(TM) system (copyright 1988, Gaussian, Inc.),

the Gaussian 86(TM) system (copyright 1986, Carnegie Mellon University), and the Gaussian 82(TM) system (copyright 1983, Carnegie Mellon University). Gaussian is a federally registered trademark of Gaussian, Inc.

This software contains proprietary and confidential information, including trade secrets, belonging to Gaussian, Inc.

This software is provided under written license and may be used, copied, transmitted, or stored only in accord with that written license.

The following legend is applicable only to US Government contracts under FAR:

RESTRICTED RIGHTS LEGEND

Use, reproduction and disclosure by the US Government is

subject to restrictions as set forth in subparagraphs (a)

and (c) of the Commercial Computer Software - Restricted

Rights clause in FAR 52.227-19.

Gaussian, Inc.

340 Quinnipiac St., Bldg. 40, Wallingford CT 06492

Warning -- This program may not be used in any manner that

competes with the business of Gaussian, Inc. or will provide

assistance to any competitor of Gaussian, Inc. The licensee

of this program is prohibited from giving any competitor of

Gaussian, Inc. access to this program. By using this program,

the user acknowledges that Gaussian, Inc. is engaged in the

business of creating and licensing software in the field of

computational chemistry and represents and warrants to the

licensee that it is not a competitor of Gaussian, Inc. and that

it will not use this program in any manner prohibited above.

Cite this work as:

Gaussian 09, Revision B.01,

M. J. Frisch, G. W. Trucks, H. B. Schlegel, G. E. Scuseria,

M. A. Robb, J. R. Cheeseman, G. Scalmani, V. Barone, B. Mennucci,

G. A. Petersson, H. Nakatsuji, M. Caricato, X. Li, H. P. Hratchian,

A. F. Izmaylov, J. Bloino, G. Zheng, J. L. Sonnenberg, M. Hada,

M. Ehara, K. Toyota, R. Fukuda, J. Hasegawa, M. Ishida, T. Nakajima,

Y. Honda, O. Kitao, H. Nakai, T. Vreven, J. A. Montgomery, Jr.,

J. E. Peralta, F. Ogliaro, M. Bearpark, J. J. Heyd, E. Brothers,

K. N. Kudin, V. N. Staroverov, T. Keith, R. Kobayashi, J. Normand,

K. Raghavachari, A. Rendell, J. C. Burant, S. S. Iyengar, J. Tomasi,

M. Cossi, N. Rega, J. M. Millam, M. Klene, J. E. Knox, J. B. Cross,

V. Bakken, C. Adamo, J. Jaramillo, R. Gomperts, R. E. Stratmann,

O. Yazyev, A. J. Austin, R. Cammi, C. Pomelli, J. W. Ochterski,

R. L. Martin, K. Morokuma, V. G. Zakrzewski, G. A. Voth,

P. Salvador, J. J. Dannenberg, S. Dapprich, A. D. Daniels,

O. Farkas, J. B. Foresman, J. V. Ortiz, J. Cioslowski,

and D. J. Fox, Gaussian, Inc., Wallingford CT, 2010.

****

Gaussian 09: IA32W-G09RevB.01 12-Aug-2010

29-Nov-2020

****

%chk=C:\Users\hp\Desktop\ch4.chk

----

opt freq b3lyp/6-311+g(d) geom=connectivity

1/14=-1,18=20,19=15,26=3,38=1,57=2/1,3;

2/9=110,12=2,17=6,18=5,40=1/2;

3/5=4,6=6,7=11,11=2,16=1,25=1,30=1,71=1,74=-5/1,2,3;

4//1;

5/5=2,38=5/2;

6/7=2,8=2,9=2,10=2,28=1/1;

7//1,2,3,16;

1/14=-1,18=20,19=15/3(2);

2/9=110/2;

99//99;

2/9=110/2;

3/5=4,6=6,7=11,11=2,16=1,25=1,30=1,71=1,74=-5/1,2,3;

4/5=5,16=3/1;

5/5=2,38=5/2;

7//1,2,3,16;

1/14=-1,18=20,19=15/3(-5);

2/9=110/2;

6/7=2,8=2,9=2,10=2,19=2,28=1/1;

99/9=1/99;

Title Card Required

Symbolic Z-matrix:

Charge = 0 Multiplicity = 1

C 1.76692 0.56391 0.

H 2.12357 -0.4449 0.

H 2.12359 1.06831 0.87365

H 2.12359 1.06831 -0.87365

H 0.69692 0.56392 0.

GradGradGradGradGradGradGradGradGradGradGradGradGradGradGradGradGradGrad

Berny optimization.

Initialization pass.

! Initial Parameters !

! (Angstroms and Degrees) !

------------------------- -------------------------

! Name Definition Value Derivative Info. !

! R1 R(1,2) 1.07 estimate D2E/DX2 !

! R2 R(1,3) 1.07 estimate D2E/DX2 !

! R3 R(1,4) 1.07 estimate D2E/DX2 !

! R4 R(1,5) 1.07 estimate D2E/DX2 !.

! A1 A(2,1,3) 109.4712 estimate D2E/DX2 !

! A2 A(2,1,4) 109.4712 estimate D2E/DX2 !

! A3 A(2,1,5) 109.4712 estimate D2E/DX2 !

! A4 A(3,1,4) 109.4713 estimate D2E/DX2 !

! A5 A(3,1,5) 109.4712 estimate D2E/DX2 !

! A6 A(4,1,5) 109.4712 estimate D2E/DX2 !

! D1 D(2,1,4,3) -120.0 estimate D2E/DX2 !

! D2 D(2,1,5,3) 120.0 estimate D2E/DX2 !

! D3 D(2,1,5,4) -120.0 estimate D2E/DX2 !

! D4 D(3,1,5,4) 120.0 estimate D2E/DX2 !

Trust Radius=3.00D-01 FncErr=1.00D-07 GrdErr=1.00D-06

Number of steps in this run= 24 maximum allowed number of steps= 100.

GradGradGradGradGradGradGradGradGradGradGradGradGradGradGradGradGradGrad

Input orientation:

Center Atomic Atomic Coordinates (Angstroms)

Number Number Type X Y Z

----

1 6 0 1.766917 0.563910 0.000000

2 1 0 2.123572 -0.444900 0.000000

3 1 0 2.123590 1.068308 0.873652

4 1 0 2.123590 1.068308 -0.873652

5 1 0 0.696917 0.563923 0.000000

Distance matrix (angstroms):

1 2 3 4 5

1 C 0.000000

2 H 1.070000 0.000000

3 H 1.070000 1.747302 0.000000

4 H 1.070000 1.747302 1.747303 0.000000

5 H 1.070000 1.747303 1.747303 1.747303 0.000000

Stoichiometry CH_4

Framework group T[O(C),4C3(H)]

Deg. of freedom 1

Full point group T NOp 12

Largest Abelian subgroup D2 NOp 4

Largest concise Abelian subgroup D2 NOp 4

Standard orientation:

Center Atomic Atomic Coordinates (Angstroms)

Number Number Type X Y Z

----

1 6 0 0.000000 0.000000 0.000000

2 1 0 0.617765 0.617765 0.617765

3 1 0 -0.617765 -0.617765 0.617765

4 1 0 -0.617765 0.617765 -0.617765

5 1 0 0.617765 -0.617765 -0.617765

Rotational constants (GHZ): 164.2463768 164.2463768 164.2463768

Standard basis: 6-311+G(d) (5D, 7F)

There are 10 symmetry adapted basis functions of A symmetry.

There are 8 symmetry adapted basis functions of B1 symmetry.

There are 8 symmetry adapted basis functions of B2 symmetry.

There are 8 symmetry adapted basis functions of B3 symmetry.

Integral buffers will be 262144 words long.

Raffenetti 2 integral format.

Two-electron integral symmetry is turned on.

34 basis functions, 56 primitive gaussians, 35 cartesian basis functions

5 alpha electrons 5 beta electrons

nuclear repulsion energy 13.6865184815 Hartrees.

NAtoms= 5 NActive= 5 NUniq= 2 SFac= 4.00D+00 NAtFMM= 50 NAOKFM=F Big=F

One-electron integrals computed using PRISM.

NBasis= 34 RedAO= T NBF= 10 8 8 8

NBsUse= 34 1.00D-06 NBFU= 10 8 8 8

Harris functional with IExCor= 402 diagonalized for initial guess.

ExpMin= 4.38D-02 ExpMax= 4.56D+03 ExpMxC= 6.82D+02 IAcc=2 IRadAn= 0 AccDes= 0.00D+00

HarFok: IExCor= 402 AccDes= 0.00D+00 IRadAn= 0 IDoV= 1

ScaDFX= 1.000000 1.000000 1.000000 1.000000

FoFCou: FMM=F IPFlag= 0 FMFlag= 100000 FMFlg1= 0

NFxFlg= 0 DoJE=T BraDBF=F KetDBF=T FulRan=T

Omega= 0.000000 0.000000 1.000000 0.000000 0.000000 ICntrl= 500 IOpCl= 0

NMat0= 1 NMatS0= 1 NMatT0= 0 NMatD0= 1 NMtDS0= 0 NMtDT0= 0

I1Cent= 4 NGrid= 0.

Petite list used in FoFCou.

Initial guess orbital symmetries:

Occupied (A) (A) (T) (T) (T)

Virtual (A) (T) (T) (T) (T) (T) (T) (A) (T) (T) (T) (T)

(T) (T) (A) (A) (E) (E) (T) (T) (T) (A) (T) (T)

(T) (T) (T) (T) (A)

The electronic state of the initial guess is 1-A.

Requested convergence on RMS density matrix=1.00D-08 within 128 cycles.

Requested convergence on MAX density matrix=1.00D-06.

Requested convergence on energy=1.00D-06.

No special actions if energy rises.

Keep R1 ints in memory in canonical form, NReq=1118393.

SCF Done: E(RB3LYP) = -40.5269656822 A.U. after 7 cycles

Convg = 0.1651D-08 -V/T = 2.0025

****

Population analysis using the SCF density.

****

Orbital symmetries:

Occupied (A) (A) (T) (T) (T)

Virtual (A) (T) (T) (T) (T) (T) (T) (A) (T) (T) (T) (T)

(T) (T) (A) (A) (E) (E) (T) (T) (T) (A) (T) (T)

(T) (T) (T) (T) (A)

The electronic state is 1-A.

Alpha occ. eigenvalues -- -10.14229 -0.70375 -0.39971 -0.39971 -0.39971

Alpha virt. eigenvalues -- 0.01085 0.05743 0.05743 0.05743 0.15177

Alpha virt. eigenvalues -- 0.15177 0.15177 0.21540 0.40328 0.40328

Alpha virt. eigenvalues -- 0.40328 0.60964 0.60964 0.60964 0.61552

Alpha virt. eigenvalues -- 0.87392 1.25394 1.25394 1.59531 1.59531

Alpha virt. eigenvalues -- 1.59531 2.34714 2.44084 2.44084 2.44084

Alpha virt. eigenvalues -- 3.13406 3.13406 3.13406 23.45834

Condensed to atoms (all electrons):

1 2 3 4 5

1 C 5.438146 0.375862 0.375862 0.375862 0.375862

2 H 0.375862 0.471076 -0.027446 -0.027446 -0.027446

3 H 0.375862 -0.027446 0.471076 -0.027446 -0.027446

4 H 0.375862 -0.027446 -0.027446 0.471076 -0.027446

5 H 0.375862 -0.027446 -0.027446 -0.027446 0.471076

Mulliken atomic charges:

1

1 C -0.941595

2 H 0.235399

3 H 0.235399

4 H 0.235399

5 H 0.235399

Sum of Mulliken atomic charges = 0.00000

Mulliken charges with hydrogens summed into heavy atoms:

1

1 C 0.000000

Sum of Mulliken charges with hydrogens summed into heavy atoms = 0.00000

Electronic spatial extent (au): <R**2>= 35.2014

Charge= 0.0000 electrons

Dipole moment (field-independent basis, Debye):

X= 0.0000 Y= 0.0000 Z= 0.0000 Tot= 0.0000

Quadrupole moment (field-independent basis, Debye-Ang):

XX= -8.4501 YY= -8.4501 ZZ= -8.4501

XY= 0.0000 XZ= 0.0000 YZ= 0.0000

Traceless Quadrupole moment (field-independent basis, Debye-Ang):

XX= 0.0000 YY= 0.0000 ZZ= 0.0000

XY= 0.0000 XZ= 0.0000 YZ= 0.0000

Octapole moment (field-independent basis, Debye-Ang**2):

XXX= 0.0000 YYY= 0.0000 ZZZ= 0.0000 XYY= 0.0000

XXY= 0.0000 XXZ= 0.0000 XZZ= 0.0000 YZZ= 0.0000

YYZ= 0.0000 XYZ= 0.7305

Hexadecapole moment (field-independent basis, Debye-Ang**3):

XXXX= -16.6677 YYYY= -16.6677 ZZZZ= -16.6677 XXXY= 0.0000

XXXZ= 0.0000 YYYX= 0.0000 YYYZ= 0.0000 ZZZX= 0.0000

ZZZY= 0.0000 XXYY= -5.0946 XXZZ= -5.0946 YYZZ= -5.0946

XXYZ= 0.0000 YYXZ= 0.0000 ZZXY= 0.0000

N-N= 1.368651848151D+01 E-N=-1.207227515305D+02 KE= 4.042785986351D+01

Symmetry A KE= 3.449759076308D+01

Symmetry B1 KE= 1.976756366809D+00

Symmetry B2 KE= 1.976756366809D+00

Symmetry B3 KE= 1.976756366809D+00

Calling FoFJK, ICntrl= 2127 FMM=F ISym2X=1 I1Cent= 0 IOpClX= 0 NMat=1 NMatS=1 NMatT=0.

***** Axes restored to original set *****

Center Atomic Forces (Hartrees/Bohr)

Number Number X Y Z

----

1 6 0.000000000 0.000000000 0.000000000

2 1 0.004838637 -0.013686260 -0.000000012

3 1 0.004838896 0.006843035 0.011852598

4 1 0.004838876 0.006843049 -0.011852598

5 1 -0.014516409 0.000000176 0.000000012

Cartesian Forces: Max 0.014516409 RMS 0.007496241

GradGradGradGradGradGradGradGradGradGradGradGradGradGradGradGradGradGrad

Berny optimization.

Internal Forces: Max 0.014516409 RMS 0.007759347

Search for a local minimum.

Step number 1 out of a maximum of 24

All quantities printed in internal units (Hartrees-Bohrs-Radians)

Mixed Optimization -- RFO/linear search

Second derivative matrix not updated -- first step.

The second derivative matrix:

R1 R2 R3 R4 A1

R1 0.37230

R2 0.00000 0.37230

R3 0.00000 0.00000 0.37230

R4 0.00000 0.00000 0.00000 0.37230

A1 0.00000 0.00000 0.00000 0.00000 0.16000

A2 0.00000 0.00000 0.00000 0.00000 0.00000

A3 0.00000 0.00000 0.00000 0.00000 0.00000

A4 0.00000 0.00000 0.00000 0.00000 0.00000

A5 0.00000 0.00000 0.00000 0.00000 0.00000

A6 0.00000 0.00000 0.00000 0.00000 0.00000

D1 0.00000 0.00000 0.00000 0.00000 0.00000

D2 0.00000 0.00000 0.00000 0.00000 0.00000

D3 0.00000 0.00000 0.00000 0.00000 0.00000

D4 0.00000 0.00000 0.00000 0.00000 0.00000

A2 A3 A4 A5 A6

A2 0.16000

A3 0.00000 0.16000

A4 0.00000 0.00000 0.16000

A5 0.00000 0.00000 0.00000 0.16000

A6 0.00000 0.00000 0.00000 0.00000 0.16000

D1 0.00000 0.00000 0.00000 0.00000 0.00000

D2 0.00000 0.00000 0.00000 0.00000 0.00000

D3 0.00000 0.00000 0.00000 0.00000 0.00000

D4 0.00000 0.00000 0.00000 0.00000 0.00000

D1 D2 D3 D4

D1 0.00499

D2 0.00000 0.00499

D3 0.00000 0.00000 0.00499

D4 0.00000 0.00000 0.00000 0.00499

ITU= 0

Eigenvalues --- 0.05269 0.05891 0.08766 0.16000 0.16000

Eigenvalues --- 0.37230 0.37230 0.37230 0.37230

RFO step: Lambda=-2.25043714D-03 EMin= 5.26881002D-02

Linear search not attempted -- first point.

Iteration 1 RMS(Cart)= 0.02071637 RMS(Int)= 0.00000000

Iteration 2 RMS(Cart)= 0.00000000 RMS(Int)= 0.00000000

ClnCor: largest displacement from symmetrization is 4.65D-14 for atom 3.

Variable Old X -DE/DX Delta X Delta X Delta X New X

(Linear) (Quad) (Total)

R1 2.02201 0.01452 0.00000 0.03876 0.03876 2.06076

R2 2.02201 0.01452 0.00000 0.03876 0.03876 2.06076

R3 2.02201 0.01452 0.00000 0.03876 0.03876 2.06076

R4 2.02201 0.01452 0.00000 0.03876 0.03876 2.06076

A1 1.91063 0.00000 0.00000 0.00000 0.00000 1.91063

A2 1.91063 0.00000 0.00000 0.00000 0.00000 1.91063

A3 1.91063 0.00000 0.00000 0.00000 0.00000 1.91063

A4 1.91063 0.00000 0.00000 0.00000 0.00000 1.91063

A5 1.91063 0.00000 0.00000 0.00000 0.00000 1.91063

A6 1.91063 0.00000 0.00000 0.00000 0.00000 1.91063

D1 -2.09440 0.00000 0.00000 0.00000 0.00000 -2.09440

D2 2.09440 0.00000 0.00000 0.00000 0.00000 2.09440

D3 -2.09440 0.00000 0.00000 0.00000 0.00000 -2.09440

D4 2.09440 0.00000 0.00000 0.00000 0.00000 2.09440

Item Value Threshold Converged?

Maximum Force 0.014516 0.000450 NO

RMS Force 0.007759 0.000300 NO

Maximum Displacement 0.038757 0.001800 NO

RMS Displacement 0.020716 0.001200 NO

Predicted change in Energy=-1.131979D-03

GradGradGradGradGradGradGradGradGradGradGradGradGradGradGradGradGradGrad

Input orientation:

Center Atomic Atomic Coordinates (Angstroms)

Number Number Type X Y Z

----

1 6 0 1.766917 0.563910 0.000000

2 1 0 2.130408 -0.464237 -0.000001

3 1 0 2.130427 1.077976 0.890397

4 1 0 2.130426 1.077977 -0.890397

5 1 0 0.676408 0.563923 0.000001

Distance matrix (angstroms):

1 2 3 4 5

1 C 0.000000

2 H 1.090509 0.000000

3 H 1.090509 1.780794 0.000000

4 H 1.090509 1.780794 1.780794 0.000000

5 H 1.090509 1.780794 1.780794 1.780794 0.000000

Stoichiometry CH4

Framework group TD[O(C),4C3(H)]

Deg. of freedom 1

Full point group TD NOp 24

Omega: Change in point group or standard orientation.

Old FWG=T [O(C1),4C3(H1)]

New FWG=TD [O(C1),4C3(H1)]

Largest Abelian subgroup D2 NOp 4

Largest concise Abelian subgroup D2 NOp 4

Standard orientation:

Center Atomic Atomic Coordinates (Angstroms)

Number Number Type X Y Z

----

1 6 0 0.000000 0.000000 0.000000

2 1 0 0.629606 0.629606 0.629606

3 1 0 -0.629606 -0.629606 0.629606

4 1 0 -0.629606 0.629606 -0.629606

5 1 0 0.629606 -0.629606 -0.629606

Rotational constants (GHZ): 158.1265094 158.1265094 158.1265094

Standard basis: 6-311+G(d) (5D, 7F)

There are 10 symmetry adapted basis functions of A symmetry.

There are 8 symmetry adapted basis functions of B1 symmetry.

There are 8 symmetry adapted basis functions of B2 symmetry.

There are 8 symmetry adapted basis functions of B3 symmetry.

Integral buffers will be 262144 words long.

Raffenetti 2 integral format.

Two-electron integral symmetry is turned on.

34 basis functions, 56 primitive gaussians, 35 cartesian basis functions

5 alpha electrons 5 beta electrons

nuclear repulsion energy 13.4291161863 Hartrees.

NAtoms= 5 NActive= 5 NUniq= 2 SFac= 4.00D+00 NAtFMM= 50 NAOKFM=F Big=F

One-electron integrals computed using PRISM.

NBasis= 34 RedAO= T NBF= 10 8 8 8

NBsUse= 34 1.00D-06 NBFU= 10 8 8 8

Initial guess read from the read-write file.

B after Tr= 0.000000 0.000000 0.000000

Rot= 1.000000 0.000000 0.000000 0.000000 Ang= 0.00 deg.

Initial guess orbital symmetries:

Occupied (A1) (A1) (T2) (T2) (T2)

Virtual (A1) (T2) (T2) (T2) (T2) (T2) (T2) (A1) (T2) (T2)

(T2) (T2) (T2) (T2) (A1) (A1) (E) (E) (T2) (T2)

(T2) (A1) (T2) (T2) (T2) (T2) (T2) (T2) (A1)

Harris functional with IExCor= 402 diagonalized for initial guess.

ExpMin= 4.38D-02 ExpMax= 4.56D+03 ExpMxC= 6.82D+02 IAcc=2 IRadAn= 0 AccDes= 0.00D+00

HarFok: IExCor= 402 AccDes= 0.00D+00 IRadAn= 0 IDoV= 1

ScaDFX= 1.000000 1.000000 1.000000 1.000000

FoFCou: FMM=F IPFlag= 0 FMFlag= 100000 FMFlg1= 0

NFxFlg= 0 DoJE=T BraDBF=F KetDBF=T FulRan=T

Omega= 0.000000 0.000000 1.000000 0.000000 0.000000 0.000000 ICntrl= 500 IOpCl= 0

NMat0= 1 NMatS0= 1 NMatT0= 0 NMatD0= 1 NMtDS0= 0 NMtDT0= 0

I1Cent= 4 NGrid= 0.

Petite list used in FoFCou.

Requested convergence on RMS density matrix=1.00D-08 within 128 cycles.

Requested convergence on MAX density matrix=1.00D-06.

Requested convergence on energy=1.00D-06.

No special actions if energy rises.

Keep R1 ints in memory in canonical form, NReq=1119085.

SCF Done: E(RB3LYP) = -40.5280818642 A.U. after 6 cycles

Convg = 0.6849D-08 -V/T = 2.0052

Calling FoFJK, ICntrl= 2127 FMM=F ISym2X=1 I1Cent= 0 IOpClX= 0 NMat=1 NMatS=1 NMatT=0.

***** Axes restored to original set *****

Center Atomic Forces (Hartrees/Bohr)

Number Number X Y Z

----

1 6 0.000000000 0.000000000 0.000000000

2 1 0.000050603 -0.000143133 0.000000000

3 1 0.000050606 0.000071565 0.000123956

4 1 0.000050606 0.000071565 -0.000123956

5 1 -0.000151814 0.000000002 0.000000000

Cartesian Forces: Max 0.000151814 RMS 0.000078397

GradGradGradGradGradGradGradGradGradGradGradGradGradGradGradGradGradGrad

Berny optimization.

Using GEDIIS/GDIIS optimizer.

Internal Forces: Max 0.000151814 RMS 0.000081148

Search for a local minimum.

Step number 2 out of a maximum of 24

All quantities printed in internal units (Hartrees-Bohrs-Radians)

Mixed Optimization -- En-DIIS/RFO-DIIS

Update second derivatives using D2CorX and points 1 2

DE= -1.12D-03 DEPred=-1.13D-03 R= 9.86D-01

SS= 1.41D+00 RLast= 7.75D-02 DXNew= 5.0454D-01 2.3254D-01

Trust test= 9.86D-01 RLast= 7.75D-02 DXMaxT set to 3.00D-01

The second derivative matrix:

R1 R2 R3 R4 A1

R1 0.37188

R2 -0.00042 0.37188

R3 -0.00042 -0.00042 0.37188

R4 -0.00042 -0.00042 -0.00042 0.37188

A1 0.00000 0.00000 0.00000 0.00000 0.16000

A2 0.00000 0.00000 0.00000 0.00000 0.00000

A3 0.00000 0.00000 0.00000 0.00000 0.00000

A4 0.00000 0.00000 0.00000 0.00000 0.00000

A5 0.00000 0.00000 0.00000 0.00000 0.00000

A6 0.00000 0.00000 0.00000 0.00000 0.00000

D1 0.00000 0.00000 0.00000 0.00000 0.00000

D2 0.00000 0.00000 0.00000 0.00000 0.00000

D3 0.00000 0.00000 0.00000 0.00000 0.00000

D4 0.00000 0.00000 0.00000 0.00000 0.00000

A2 A3 A4 A5 A6

A2 0.16000

A3 0.00000 0.16000

A4 0.00000 0.00000 0.16000

A5 0.00000 0.00000 0.00000 0.16000

A6 0.00000 0.00000 0.00000 0.00000 0.16000

D1 0.00000 0.00000 0.00000 0.00000 0.00000

D2 0.00000 0.00000 0.00000 0.00000 0.00000

D3 0.00000 0.00000 0.00000 0.00000 0.00000

D4 0.00000 0.00000 0.00000 0.00000 0.00000

D1 D2 D3 D4

D1 0.00499

D2 0.00000 0.00499

D3 0.00000 0.00000 0.00499

D4 0.00000 0.00000 0.00000 0.00499

ITU= 1 0

Use linear search instead of GDIIS.

Eigenvalues --- 0.05269 0.05891 0.08766 0.16000 0.16000

Eigenvalues --- 0.37063 0.37230 0.37230 0.37230

RFO step: Lambda= 0.00000000D+00 EMin= 5.26881002D-02

Quartic linear search produced a step of 0.01120.

Iteration 1 RMS(Cart)= 0.00023197 RMS(Int)= 0.00000000

Iteration 2 RMS(Cart)= 0.00000000 RMS(Int)= 0.00000000

ClnCor: largest displacement from symmetrization is 1.12D-14 for atom 4.

Variable Old X -DE/DX Delta X Delta X Delta X New X

(Linear) (Quad) (Total)

R1 2.06076 0.00015 0.00043 0.00000 0.00043 2.06120

R2 2.06076 0.00015 0.00043 0.00000 0.00043 2.06120

R3 2.06076 0.00015 0.00043 0.00000 0.00043 2.06120

R4 2.06076 0.00015 0.00043 0.00000 0.00043 2.06120

A1 1.91063 0.00000 0.00000 0.00000 0.00000 1.91063

A2 1.91063 0.00000 0.00000 0.00000 0.00000 1.91063

A3 1.91063 0.00000 0.00000 0.00000 0.00000 1.91063

A4 1.91063 0.00000 0.00000 0.00000 0.00000 1.91063

A5 1.91063 0.00000 0.00000 0.00000 0.00000 1.91063

A6 1.91063 0.00000 0.00000 0.00000 0.00000 1.91063

D1 -2.09440 0.00000 0.00000 0.00000 0.00000 -2.09440

D2 2.09440 0.00000 0.00000 0.00000 0.00000 2.09440

D3 -2.09440 0.00000 0.00000 0.00000 0.00000 -2.09440

D4 2.09440 0.00000 0.00000 0.00000 0.00000 2.09440

Item Value Threshold Converged?

Maximum Force 0.000152 0.000450 YES

RMS Force 0.000081 0.000300 YES

Maximum Displacement 0.000434 0.001800 YES

RMS Displacement 0.000232 0.001200 YES

Predicted change in Energy=-1.239284D-07

Optimization completed.

-- Stationary point found.

! Optimized Parameters !

! (Angstroms and Degrees) !

------------------------- -------------------------

! Name Definition Value Derivative Info. !

! R1 R(1,2) 1.0905 -DE/DX = 0.0002 !

! R2 R(1,3) 1.0905 -DE/DX = 0.0002 !

! R3 R(1,4) 1.0905 -DE/DX = 0.0002 !

! R4 R(1,5) 1.0905 -DE/DX = 0.0002 !

! A1 A(2,1,3) 109.4712 -DE/DX = 0.0 !

! A2 A(2,1,4) 109.4712 -DE/DX = 0.0 !

! A3 A(2,1,5) 109.4712 -DE/DX = 0.0 !

! A4 A(3,1,4) 109.4712 -DE/DX = 0.0 !

! A5 A(3,1,5) 109.4712 -DE/DX = 0.0 !

! A6 A(4,1,5) 109.4712 -DE/DX = 0.0 !

! D1 D(2,1,4,3) -120.0 -DE/DX = 0.0 !

! D2 D(2,1,5,3) 120.0 -DE/DX = 0.0 !

! D3 D(2,1,5,4) -120.0 -DE/DX = 0.0 !

! D4 D(3,1,5,4) 120.0 -DE/DX = 0.0 !

GradGradGradGradGradGradGradGradGradGradGradGradGradGradGradGradGradGrad

Input orientation:

Center Atomic Atomic Coordinates (Angstroms)

Number Number Type X Y Z

----

1 6 0 1.766917 0.563910 0.000000

2 1 0 2.130408 -0.464237 -0.000001

3 1 0 2.130427 1.077976 0.890397

4 1 0 2.130426 1.077977 -0.890397

5 1 0 0.676408 0.563923 0.000001

Distance matrix (angstroms):

1 2 3 4 5

1 C 0.000000

2 H 1.090509 0.000000

3 H 1.090509 1.780794 0.000000

4 H 1.090509 1.780794 1.780794 0.000000

5 H 1.090509 1.780794 1.780794 1.780794 0.000000

Stoichiometry CH4

Framework group TD[O(C),4C3(H)]

Deg. of freedom 1

Full point group TD NOp 24

Largest Abelian subgroup D2 NOp 4

Largest concise Abelian subgroup D2 NOp 4

Standard orientation:

Center Atomic Atomic Coordinates (Angstroms)

Number Number Type X Y Z

----

1 6 0 0.000000 0.000000 0.000000

2 1 0 0.629606 0.629606 0.629606

3 1 0 -0.629606 -0.629606 0.629606

4 1 0 -0.629606 0.629606 -0.629606

5 1 0 0.629606 -0.629606 -0.629606

Rotational constants (GHZ): 158.1265094 158.1265094 158.1265094

****

Population analysis using the SCF density.

****

Orbital symmetries:

Occupied (A1) (A1) (T2) (T2) (T2)

Virtual (A1) (T2) (T2) (T2) (T2) (T2) (T2) (A1) (T2) (T2)

(T2) (T2) (T2) (T2) (A1) (A1) (E) (E) (T2) (T2)

(T2) (A1) (T2) (T2) (T2) (T2) (T2) (T2) (A1)

The electronic state is 1-A1.

Alpha occ. eigenvalues -- -10.15133 -0.69745 -0.39634 -0.39634 -0.39634

Alpha virt. eigenvalues -- 0.00991 0.05736 0.05736 0.05736 0.14637

Alpha virt. eigenvalues -- 0.14637 0.14637 0.21072 0.40418 0.40418

Alpha virt. eigenvalues -- 0.40418 0.59906 0.59906 0.59906 0.60278

Alpha virt. eigenvalues -- 0.86494 1.25797 1.25797 1.59542 1.59542

Alpha virt. eigenvalues -- 1.59542 2.35038 2.42329 2.42329 2.42329

Alpha virt. eigenvalues -- 3.10057 3.10057 3.10057 23.43988

Condensed to atoms (all electrons):

1 2 3 4 5

1 C 5.441470 0.372433 0.372433 0.372433 0.372433

2 H 0.372433 0.475956 -0.027063 -0.027063 -0.027063

3 H 0.372433 -0.027063 0.475956 -0.027063 -0.027063

4 H 0.372433 -0.027063 -0.027063 0.475956 -0.027063

5 H 0.372433 -0.027063 -0.027063 -0.027063 0.475956

Mulliken atomic charges:

1

1 C -0.931204

2 H 0.232801

3 H 0.232801

4 H 0.232801

5 H 0.232801

Sum of Mulliken atomic charges = 0.00000

Mulliken charges with hydrogens summed into heavy atoms:

1

1 C 0.000000

Sum of Mulliken charges with hydrogens summed into heavy atoms = 0.00000

Electronic spatial extent (au): <R**2>= 35.9778

Charge= 0.0000 electrons

Dipole moment (field-independent basis, Debye):

X= 0.0000 Y= 0.0000 Z= 0.0000 Tot= 0.0000

Quadrupole moment (field-independent basis, Debye-Ang):

XX= -8.5144 YY= -8.5144 ZZ= -8.5144

XY= 0.0000 XZ= 0.0000 YZ= 0.0000

Traceless Quadrupole moment (field-independent basis, Debye-Ang):

XX= 0.0000 YY= 0.0000 ZZ= 0.0000

XY= 0.0000 XZ= 0.0000 YZ= 0.0000

Octapole moment (field-independent basis, Debye-Ang**2):

XXX= 0.0000 YYY= 0.0000 ZZZ= 0.0000 XYY= 0.0000

XXY= 0.0000 XXZ= 0.0000 XZZ= 0.0000 YZZ= 0.0000

YYZ= 0.0000 XYZ= 0.7811

Hexadecapole moment (field-independent basis, Debye-Ang**3):

XXXX= -17.2201 YYYY= -17.2201 ZZZZ= -17.2201 XXXY= 0.0000

XXXZ= 0.0000 YYYX= 0.0000 YYYZ= 0.0000 ZZZX= 0.0000

ZZZY= 0.0000 XXYY= -5.2429 XXZZ= -5.2429 YYZZ= -5.2429

XXYZ= 0.0000 YYXZ= 0.0000 ZZXY= 0.0000

N-N= 1.342911618634D+01 E-N=-1.201307897495D+02 KE= 4.031684346614D+01

Symmetry A KE= 3.447981936062D+01

Symmetry B1 KE= 1.945674701840D+00

Symmetry B2 KE= 1.945674701840D+00

Symmetry B3 KE= 1.945674701840D+00

1|1|UNPC-LAPTOP-VBAQNH7H|FOpt|RB3LYP|6-311+G(d)|C1H4|HP|29-Nov-2020|0|

|# opt freq b3lyp/6-311+g(d) geom=connectivity||Title Card Required||0

,1|C,1.76691739,0.563909725,0.|H,2.1304079871,-0.4642365757,-0.0000009

202|H,2.1304274232,1.077975738,0.8903970013|H,2.1304259205,1.077976800

6,-0.8903970013|H,0.6764082293,0.5639229371,0.0000009202||Version=IA32

W-G09RevB.01|State=1-A1|HF=-40.5280819|RMSD=6.849e-009|RMSF=7.840e-005

|Dipole=0.,0.,0.|Quadrupole=0.,0.,0.,0.,0.,0.|PG=TD [O(C1),4C3(H1)]||@

WERE I TO AWAIT PERFECTION, MY BOOK WOULD NEVER BE FINISHED.

-- HISTORY OF CHINESE WRITING

TAI T'UNG, 13TH CENTURY

Job cpu time: 0 days 0 hours 0 minutes 3.0 seconds.

File lengths (MBytes): RWF= 5 Int= 0 D2E= 0 Chk= 1 Scr= 1

Normal termination of Gaussian 09 at Sun Nov 29 23:31:54 2020.

Link1: Proceeding to internal job step number 2.

#N Geom=AllCheck Guess=TCheck SCRF=Check GenChk RB3LYP/6-311+G(d) Freq

----

1/10=4,29=7,30=1,38=1,40=1/1,3;

2/12=2,40=1/2;

3/5=4,6=6,7=11,11=2,14=-4,16=1,25=1,30=1,70=2,71=2,74=-5,116=1/1,2,3;

4/5=101/1;

5/5=2,98=1/2;

8/6=4,10=90,11=11/1;

11/6=1,8=1,9=11,15=111,16=1/1,2,10;

10/6=1/2;

6/7=2,8=2,9=2,10=2,18=1,28=1/1;

7/8=1,10=1,25=1/1,2,3,16;

1/10=4,30=1/3;

99//99;

Title Card Required

Redundant internal coordinates taken from checkpoint file:

C:\Users\hp\Desktop\ch4.chk

Charge = 0 Multiplicity = 1

C,0,1.76691739,0.563909725,0.

H,0,2.1304079871,-0.4642365757,-0.0000009202

H,0,2.1304274232,1.077975738,0.8903970013

H,0,2.1304259205,1.0779768006,-0.8903970013

H,0,0.6764082293,0.5639229371,0.0000009202

Recover connectivity data from disk.

GradGradGradGradGradGradGradGradGradGradGradGradGradGradGradGradGradGrad

Berny optimization.

Initialization pass.

! Initial Parameters !

! (Angstroms and Degrees) !

------------------------- -------------------------

! Name Definition Value Derivative Info. !

! R1 R(1,2) 1.0905 calculate D2E/DX2 analytically !

! R2 R(1,3) 1.0905 calculate D2E/DX2 analytically !

! R3 R(1,4) 1.0905 calculate D2E/DX2 analytically !

! R4 R(1,5) 1.0905 calculate D2E/DX2 analytically !

! A1 A(2,1,3) 109.4712 calculate D2E/DX2 analytically !

! A2 A(2,1,4) 109.4712 calculate D2E/DX2 analytically !

! A3 A(2,1,5) 109.4712 calculate D2E/DX2 analytically !

! A4 A(3,1,4) 109.4712 calculate D2E/DX2 analytically !

! A5 A(3,1,5) 109.4712 calculate D2E/DX2 analytically !

! A6 A(4,1,5) 109.4712 calculate D2E/DX2 analytically !

! D1 D(2,1,4,3) -120.0 calculate D2E/DX2 analytically !

! D2 D(2,1,5,3) 120.0 calculate D2E/DX2 analytically !

! D3 D(2,1,5,4) -120.0 calculate D2E/DX2 analytically !

! D4 D(3,1,5,4) 120.0 calculate D2E/DX2 analytically !

Trust Radius=3.00D-01 FncErr=1.00D-07 GrdErr=1.00D-07

Number of steps in this run= 2 maximum allowed number of steps= 2.

GradGradGradGradGradGradGradGradGradGradGradGradGradGradGradGradGradGrad

Input orientation:

Center Atomic Atomic Coordinates (Angstroms)

Number Number Type X Y Z

----

1 6 0 1.766917 0.563910 0.000000

2 1 0 2.130408 -0.464237 -0.000001

3 1 0 2.130427 1.077976 0.890397

4 1 0 2.130426 1.077977 -0.890397

5 1 0 0.676408 0.563923 0.000001

Distance matrix (angstroms):

1 2 3 4 5

1 C 0.000000

2 H 1.090509 0.000000

3 H 1.090509 1.780794 0.000000

4 H 1.090509 1.780794 1.780794 0.000000

5 H 1.090509 1.780794 1.780794 1.780794 0.000000

Stoichiometry CH4

Framework group TD[O(C),4C3(H)]

Deg. of freedom 1

Full point group TD NOp 24

Largest Abelian subgroup D2 NOp 4

Largest concise Abelian subgroup D2 NOp 4

Standard orientation:

Center Atomic Atomic Coordinates (Angstroms)

Number Number Type X Y Z

---

1 6 0 0.000000 0.000000 0.000000

2 1 0 0.629606 0.629606 0.629606

3 1 0 -0.629606 -0.629606 0.629606

4 1 0 -0.629606 0.629606 -0.629606

5 1 0 0.629606 -0.629606 -0.629606

Rotational constants (GHZ): 158.1265094 158.1265094 158.1265094

Standard basis: 6-311+G(d) (5D, 7F)

There are 10 symmetry adapted basis functions of A symmetry.

There are 8 symmetry adapted basis functions of B1 symmetry.

There are 8 symmetry adapted basis functions of B2 symmetry.

There are 8 symmetry adapted basis functions of B3 symmetry.

Integral buffers will be 262144 words long.

Raffenetti 2 integral format.

Two-electron integral symmetry is turned on.

34 basis functions, 56 primitive gaussians, 35 cartesian basis functions

5 alpha electrons 5 beta electrons

nuclear repulsion energy 13.4291161863 Hartrees.

NAtoms= 5 NActive= 5 NUniq= 2 SFac= 4.00D+00 NAtFMM= 50 NAOKFM=F Big=F

One-electron integrals computed using PRISM.

NBasis= 34 RedAO= T NBF= 10 8 8 8

NBsUse= 34 1.00D-06 NBFU= 10 8 8 8

Initial guess read from the checkpoint file: C:\Users\hp\Desktop\ch4.chk

B after Tr= 0.000000 0.000000 0.000000

Rot= 1.000000 0.000000 0.000000 0.000000 Ang= 0.00 deg.

Initial guess orbital symmetries:

Occupied (A1) (A1) (T2) (T2) (T2)

Virtual (A1) (T2) (T2) (T2) (T2) (T2) (T2) (A1) (T2) (T2)

(T2) (T2) (T2) (T2) (A1) (A1) (E) (E) (T2) (T2)

(T2) (A1) (T2) (T2) (T2) (T2) (T2) (T2) (A1)

Requested convergence on RMS density matrix=1.00D-08 within 128 cycles.

Requested convergence on MAX density matrix=1.00D-06.

Requested convergence on energy=1.00D-06.

No special actions if energy rises.

Keep R1 ints in memory in canonical form, NReq=1119085.

SCF Done: E(RB3LYP) = -40.5280818642 A.U. after 1 cycles

Convg = 0.7046D-09 -V/T = 2.0052

Range of M.O.s used for correlation: 1 34

NBasis= 34 NAE= 5 NBE= 5 NFC= 0 NFV= 0

NROrb= 34 NOA= 5 NOB= 5 NVA= 29 NVB= 29

Symmetrizing basis deriv contribution to polar:

IMax=3 JMax=2 DiffMx= 0.00D+00

G2DrvN: will do 6 centers at a time, making 1 passes doing MaxLOS=2.

Calling FoFCou, ICntrl= 3107 FMM=F I1Cent= 0 AccDes= 0.00D+00.

FoFDir/FoFCou used for L=0 through L=2.

End of G2Drv Frequency-dependent properties file 721 does not exist.

End of G2Drv Frequency-dependent properties file 722 does not exist.

IDoAtm=11111

Differentiating once with respect to electric field.

with respect to dipole field.

Differentiating once with respect to nuclear coordinates.

Keep R1 ints in memory in canonical form, NReq=1020123.

There are 9 degrees of freedom in the 1st order CPHF. IDoFFX=4.

9 vectors produced by pass 0 Test12= 1.61D-15 1.11D-08 XBig12= 7.20D+00 1.49D+00.

AX will form 9 AO Fock derivatives at one time.

9 vectors produced by pass 1 Test12= 1.61D-15 1.11D-08 XBig12= 2.01D-01 2.19D-01.

9 vectors produced by pass 2 Test12= 1.61D-15 1.11D-08 XBig12= 6.03D-04 7.70D-03.

9 vectors produced by pass 3 Test12= 1.61D-15 1.11D-08 XBig12= 1.57D-06 4.26D-04.

9 vectors produced by pass 4 Test12= 1.61D-15 1.11D-08 XBig12= 5.58D-10 7.80D-06.

4 vectors produced by pass 5 Test12= 1.61D-15 1.11D-08 XBig12= 3.26D-13 2.44D-07.

Inverted reduced A of dimension 49 with in-core refinement.

Isotropic polarizability for W= 0.000000 13.81 Bohr**3.

End of Minotr Frequency-dependent properties file 721 does not exist.

End of Minotr Frequency-dependent properties file 722 does not exist.

****

Population analysis using the SCF density.

****

Orbital symmetries:

Occupied (A1) (A1) (T2) (T2) (T2)

Virtual (A1) (T2) (T2) (T2) (T2) (T2) (T2) (A1) (T2) (T2)

(T2) (T2) (T2) (T2) (A1) (A1) (E) (E) (T2) (T2)

(T2) (A1) (T2) (T2) (T2) (T2) (T2) (T2) (A1)

The electronic state is 1-A1.

Alpha occ. eigenvalues -- -10.15133 -0.69745 -0.39634 -0.39634 -0.39634

Alpha virt. eigenvalues -- 0.00991 0.05736 0.05736 0.05736 0.14637

Alpha virt. eigenvalues -- 0.14637 0.14637 0.21072 0.40418 0.40418

Alpha virt. eigenvalues -- 0.40418 0.59906 0.59906 0.59906 0.60278

Alpha virt. eigenvalues -- 0.86494 1.25797 1.25797 1.59542 1.59542

Alpha virt. eigenvalues -- 1.59542 2.35038 2.42329 2.42329 2.42329

Alpha virt. eigenvalues -- 3.10057 3.10057 3.10057 23.43988

Condensed to atoms (all electrons):

1 2 3 4 5

1 C 5.441470 0.372433 0.372433 0.372433 0.372433

2 H 0.372433 0.475956 -0.027063 -0.027063 -0.027063

3 H 0.372433 -0.027063 0.475956 -0.027063 -0.027063

4 H 0.372433 -0.027063 -0.027063 0.475956 -0.027063

5 H 0.372433 -0.027063 -0.027063 -0.027063 0.475956

Mulliken atomic charges:

1

1 C -0.931204

2 H 0.232801

3 H 0.232801

4 H 0.232801

5 H 0.232801

Sum of Mulliken atomic charges = 0.00000

Mulliken charges with hydrogens summed into heavy atoms:

1

1 C 0.000000

Sum of Mulliken charges with hydrogens summed into heavy atoms = 0.00000

APT atomic charges:

1

1 C -0.015942

2 H 0.003985

3 H 0.003985

4 H 0.003985

5 H 0.003985

Sum of APT charges= 0.00000

APT Atomic charges with hydrogens summed into heavy atoms:

1

1 C 0.000000

2 H 0.000000

3 H 0.000000

4 H 0.000000

5 H 0.000000

Sum of APT charges= 0.00000

Electronic spatial extent (au): <R**2>= 35.9778

Charge= 0.0000 electrons

Dipole moment (field-independent basis, Debye):

X= 0.0000 Y= 0.0000 Z= 0.0000 Tot= 0.0000

Quadrupole moment (field-independent basis, Debye-Ang):

XX= -8.5144 YY= -8.5144 ZZ= -8.5144

XY= 0.0000 XZ= 0.0000 YZ= 0.0000

Traceless Quadrupole moment (field-independent basis, Debye-Ang):

XX= 0.0000 YY= 0.0000 ZZ= 0.0000

XY= 0.0000 XZ= 0.0000 YZ= 0.0000

Octapole moment (field-independent basis, Debye-Ang**2):

XXX= 0.0000 YYY= 0.0000 ZZZ= 0.0000 XYY= 0.0000

XXY= 0.0000 XXZ= 0.0000 XZZ= 0.0000 YZZ= 0.0000

YYZ= 0.0000 XYZ= 0.7811

Hexadecapole moment (field-independent basis, Debye-Ang**3):

XXXX= -17.2201 YYYY= -17.2201 ZZZZ= -17.2201 XXXY= 0.0000

XXXZ= 0.0000 YYYX= 0.0000 YYYZ= 0.0000 ZZZX= 0.0000

ZZZY= 0.0000 XXYY= -5.2429 XXZZ= -5.2429 YYZZ= -5.2429

XXYZ= 0.0000 YYXZ= 0.0000 ZZXY= 0.0000

N-N= 1.342911618634D+01 E-N=-1.201307898411D+02 KE= 4.031684350874D+01

Symmetry A KE= 3.447981938511D+01

Symmetry B1 KE= 1.945674707876D+00

Symmetry B2 KE= 1.945674707876D+00

Symmetry B3 KE= 1.945674707876D+00

Exact polarizability: 13.812 0.000 13.812 0.000 0.000 13.812

Approx polarizability: 16.443 0.000 16.443 0.000 0.000 16.443

Calling FoFJK, ICntrl= 100127 FMM=F ISym2X=1 I1Cent= 0 IOpClX= 0 NMat=1 NMatS=1 NMatT=0.

Full mass-weighted force constant matrix:

Low frequencies --- -0.0012 -0.0011 -0.0003 46.1637 46.1637 46.1637

Low frequencies --- 1355.1285 1355.1285 1355.1285

Harmonic frequencies (cm**-1), IR intensities (KM/Mole), Raman scattering

activities (A**4/AMU), depolarization ratios for plane and unpolarized

incident light, reduced masses (AMU), force constants (mDyne/A),

and normal coordinates:

1 2 3

T2 T2 T2

Frequencies -- 1355.1285 1355.1285 1355.1285

Red. masses -- 1.1788 1.1788 1.1788

Frc consts -- 1.2754 1.2754 1.2754

IR Inten -- 20.7468 20.7468 20.7468

Atom AN X Y Z X Y Z X Y Z

1 6 -0.02 0.12 0.01 0.12 0.02 0.00 0.00 -0.01 0.12

2 1 0.31 -0.39 0.16 -0.32 0.17 0.26 0.22 0.26 -0.39

3 1 0.28 -0.42 -0.22 -0.33 0.16 -0.28 -0.24 -0.21 -0.35

4 1 -0.15 -0.34 -0.30 -0.40 -0.30 0.18 0.26 -0.20 -0.36

5 1 -0.18 -0.31 0.24 -0.41 -0.29 -0.20 -0.21 0.27 -0.38

4 5 6

E E A1

Frequencies -- 1582.6386 1582.6386 3030.2510

Red. masses -- 1.0078 1.0078 1.0078

Frc consts -- 1.4873 1.4873 5.4525

IR Inten -- 0.0000 0.0000 0.0000

Atom AN X Y Z X Y Z X Y Z

1 6 0.00 0.00 0.00 0.00 0.00 0.00 0.00 0.00 0.00

2 1 -0.32 0.38 -0.07 -0.26 -0.14 0.40 -0.29 -0.29 -0.29

3 1 0.32 -0.38 -0.07 0.26 0.14 0.40 0.29 0.29 -0.29

4 1 0.32 0.38 0.07 0.26 -0.14 -0.40 0.29 -0.29 0.29

5 1 -0.32 -0.38 0.07 -0.26 0.14 -0.40 -0.29 0.29 0.29

7 8 9

T2 T2 T2

Frequencies -- 3135.0616 3135.0616 3135.0616

Red. masses -- 1.1018 1.1018 1.1018

Frc consts -- 6.3804 6.3804 6.3804

IR Inten -- 29.9080 29.9080 29.9080

Atom AN X Y Z X Y Z X Y Z

1 6 0.01 -0.09 0.00 -0.07 -0.01 0.06 -0.06 -0.01 -0.07

2 1 0.24 0.22 0.24 0.05 0.06 0.07 0.43 0.44 0.43

3 1 0.25 0.23 -0.25 0.43 0.44 -0.43 -0.01 -0.01 -0.01

4 1 -0.34 0.32 -0.34 -0.01 0.00 0.01 0.36 -0.38 0.36

5 1 -0.33 0.31 0.33 0.37 -0.38 -0.37 -0.08 0.07 0.06

- Thermochemistry -

Temperature 298.150 Kelvin. Pressure 1.00000 Atm.

Atom 1 has atomic number 6 and mass 12.00000

Atom 2 has atomic number 1 and mass 1.00783

Atom 3 has atomic number 1 and mass 1.00783

Atom 4 has atomic number 1 and mass 1.00783

Atom 5 has atomic number 1 and mass 1.00783

Molecular mass: 16.03130 amu.

Principal axes and moments of inertia in atomic units:

1 2 3

Eigenvalues -- 11.41327 11.41327 11.41327

X 0.04958 0.97439 -0.21932

Y -0.21486 0.22486 0.95041

Z 0.97539 0.00000 0.22050

This molecule is a spherical top.

Rotational symmetry number 12.

Rotational temperatures (Kelvin) 7.58887 7.58887 7.58887

Rotational constants (GHZ): 158.12651 158.12651 158.12651

Zero-point vibrational energy 117629.4 (Joules/Mol)

28.11410 (Kcal/Mol)

Vibrational temperatures: 1949.73 1949.73 1949.73 2277.06 2277.06

(Kelvin) 4359.85 4510.65 4510.65 4510.65

Zero-point correction= 0.044803 (Hartree/Particle)

Thermal correction to Energy= 0.047669

Thermal correction to Enthalpy= 0.048613

Thermal correction to Gibbs Free Energy= 0.027486

Sum of electronic and zero-point Energies= -40.483279

Sum of electronic and thermal Energies= -40.480413

Sum of electronic and thermal Enthalpies= -40.479469

Sum of electronic and thermal Free Energies= -40.500596

E (Thermal) CV S

KCal/Mol Cal/Mol-Kelvin Cal/Mol-Kelvin

Total 29.913 6.444 44.465

Electronic 0.000 0.000 0.000

Translational 0.889 2.981 34.261

Rotational 0.889 2.981 10.122

Vibrational 28.135 0.482 0.082

Q Log10(Q) Ln(Q)

Total Bot 0.227635D-12 -12.642760 -29.111032

Total V=0 0.922556D+08 7.964993 18.340074

Vib (Bot) 0.248057D-20 -20.605449 -47.445800

Vib (V=0) 0.100532D+01 0.002304 0.005305

Electronic 0.100000D+01 0.000000 0.000000

Translational 0.252295D+07 6.401908 14.740939

Rotational 0.363731D+02 1.560781 3.593830

***** Axes restored to original set *****

Center Atomic Forces (Hartrees/Bohr)

Number Number X Y Z

----

1 6 0.000000000 0.000000000 0.000000000

2 1 0.000050603 -0.000143132 0.000000000

3 1 0.000050606 0.000071565 0.000123956

4 1 0.000050605 0.000071565 -0.000123956

5 1 -0.000151814 0.000000002 0.000000000

Cartesian Forces: Max 0.000151814 RMS 0.000078397

GradGradGradGradGradGradGradGradGradGradGradGradGradGradGradGradGradGrad

Berny optimization.

Internal Forces: Max 0.000151814 RMS 0.000081148

Search for a local minimum.

Step number 1 out of a maximum of 2

All quantities printed in internal units (Hartrees-Bohrs-Radians)

Second derivative matrix not updated -- analytic derivatives used.

The second derivative matrix:

R1 R2 R3 R4 A1

R1 0.34282

R2 0.00246 0.34282

R3 0.00246 0.00246 0.34282

R4 0.00246 0.00246 0.00246 0.34282

A1 0.00188 0.00188 -0.00118 -0.00257 0.02004

A2 0.00294 -0.00076 0.00327 -0.00545 -0.00893

A3 0.00574 -0.00461 -0.00510 0.00397 -0.01329

A4 -0.00076 0.00294 0.00327 -0.00545 -0.00893

A5 -0.00461 0.00574 -0.00510 0.00397 -0.01329

A6 -0.00519 -0.00519 0.00484 0.00554 0.02442

D1 -0.00363 -0.00363 -0.00256 0.00983 -0.01360

D2 0.00299 0.00299 -0.00769 0.00171 0.00826

D3 -0.00394 0.00693 -0.00385 0.00085 0.00413

D4 -0.00693 0.00394 0.00385 -0.00085 -0.00413

A2 A3 A4 A5 A6

A2 0.03773

A3 -0.02295 0.07372

A4 -0.00218 0.02535 0.03773

A5 0.02535 -0.03581 -0.02295 0.07372

A6 -0.02902 -0.02701 -0.02902 -0.02701 0.08763

D1 -0.01083 0.01481 -0.01083 0.01481 0.00564

D2 -0.00947 0.00693 -0.00947 0.00693 -0.00317

D3 -0.01438 -0.00049 0.00491 0.00742 -0.00159

D4 -0.00491 -0.00742 0.01438 0.00049 0.00159

D1 D2 D3 D4

D1 0.02992

D2 0.00149 0.01860

D3 0.00074 0.00930 0.01889

D4 -0.00074 -0.00930 0.00959 0.01889

ITU= 0

Eigenvalues --- 0.03818 0.04300 0.06530 0.13346 0.13403

Eigenvalues --- 0.34123 0.34124 0.34150 0.35021

Angle between quadratic step and forces= 0.00 degrees.

Linear search not attempted -- first point.

Iteration 1 RMS(Cart)= 0.00023171 RMS(Int)= 0.00000000

Iteration 2 RMS(Cart)= 0.00000000 RMS(Int)= 0.00000000

ClnCor: largest displacement from symmetrization is 1.11D-12 for atom 4.

Variable Old X -DE/DX Delta X Delta X Delta X New X

(Linear) (Quad) (Total)

R1 2.06076 0.00015 0.00000 0.00043 0.00043 2.06120

R2 2.06076 0.00015 0.00000 0.00043 0.00043 2.06120

R3 2.06076 0.00015 0.00000 0.00043 0.00043 2.06120

R4 2.06076 0.00015 0.00000 0.00043 0.00043 2.06120

A1 1.91063 0.00000 0.00000 0.00000 0.00000 1.91063

A2 1.91063 0.00000 0.00000 0.00000 0.00000 1.91063

A3 1.91063 0.00000 0.00000 0.00000 0.00000 1.91063

A4 1.91063 0.00000 0.00000 0.00000 0.00000 1.91063

A5 1.91063 0.00000 0.00000 0.00000 0.00000 1.91063

A6 1.91063 0.00000 0.00000 0.00000 0.00000 1.91063

D1 -2.09440 0.00000 0.00000 0.00000 0.00000 -2.09440

D2 2.09440 0.00000 0.00000 0.00000 0.00000 2.09440

D3 -2.09440 0.00000 0.00000 0.00000 0.00000 -2.09440

D4 2.09440 0.00000 0.00000 0.00000 0.00000 2.09440

Item Value Threshold Converged?

Maximum Force 0.000152 0.000450 YES

RMS Force 0.000081 0.000300 YES

Maximum Displacement 0.000433 0.001800 YES

RMS Displacement 0.000232 0.001200 YES

Predicted change in Energy=-1.316197D-07

Optimization completed.

-- Stationary point found.

! Optimized Parameters !

! (Angstroms and Degrees) !

------------------------- -------------------------

! Name Definition Value Derivative Info. !

! R1 R(1,2) 1.0905 -DE/DX = 0.0002 !

! R2 R(1,3) 1.0905 -DE/DX = 0.0002 !

! R3 R(1,4) 1.0905 -DE/DX = 0.0002 !

! R4 R(1,5) 1.0905 -DE/DX = 0.0002 !

! A1 A(2,1,3) 109.4712 -DE/DX = 0.0 !

! A2 A(2,1,4) 109.4712 -DE/DX = 0.0 !

! A3 A(2,1,5) 109.4712 -DE/DX = 0.0 !

! A4 A(3,1,4) 109.4712 -DE/DX = 0.0 !

! A5 A(3,1,5) 109.4712 -DE/DX = 0.0 !

! A6 A(4,1,5) 109.4712 -DE/DX = 0.0 !

! D1 D(2,1,4,3) -120.0 -DE/DX = 0.0 !

! D2 D(2,1,5,3) 120.0 -DE/DX = 0.0 !

! D3 D(2,1,5,4) -120.0 -DE/DX = 0.0 !

! D4 D(3,1,5,4) 120.0 -DE/DX = 0.0 !

GradGradGradGradGradGradGradGradGradGradGradGradGradGradGradGradGradGrad

1|1|UNPC-LAPTOP-VBAQNH7H|Freq|RB3LYP|6-311+G(d)|C1H4|HP|29-Nov-2020|0|

|#N Geom=AllCheck Guess=TCheck SCRF=Check GenChk RB3LYP/6-311+G(d) Freq||Title
Card   Required||0,1|C,1.76691739,0.563909725,0.|H,2.1304079871,-0.4642365757-
-0.0000009202|H,2.1304274232,1.077975738,0.8903970013|H

,2.1304259205,1.0779768006,-
0.8903970013|H,0.6764082293,0.5639229371,0.0000009202||Version=IA32W-G09RevB.01|
State=1-A1|HF=-40.5280819|RMSD=7.046e-010|RMSF=7.8-
0e-005|ZeroPoint=0.0448027|Thermal=0.047669|Dipole=

0.,0.,0.|DipoleDeriv=-0.015942,0.,0.,0.,-0.015942,0.,0.,0.,-0.015942,0

.0578811,0.076215,0.,0.076215,-0.1307508,-0.0000002,0.,-0.0000002,0.08

48261,0.0578782,-0.038109,-0.0660074,-0.038109,0.0309334,-0.0933458,-0.066-
074,-0.0933458,-0.0768551,0.0578784,-0.0381089,0.0660071,-0.03810
89,0.0309332,0.093346,0.0660071,0.093346,-0.0768551,
0.1576958,0.0000029,0.0000002,0.0000029,0.0848261,0.,0.0000002,0.,0.0848261|Polar=13.
8120382,0.,13.8120382,0.,0.,13.8120382|PG=TD
[O(C1),4C3(H1)]|NImag=0||0.54767104,0.,0.54767104,0.,0.,0.54767104,-0.07688286,0.0848
9672,0.00000008,0.07885896,0.08489672,-0.28700193,-0.0000002-
,-0.09332226,0.30983122,0.00000008,-0.00000021,-0.046868-
9,-0.00000008,0.00000024,0.04586582,-0.0768860-
,-0.04245004,0.07352633,0.00331311,0.00442709,0.01113397

,0.07886248,-0.04245004,-0.10690019,-0.10397894,-0.01185643,-0.0127-
859,-0.02558442,0.04666298,0.11185534,-0.07352633,-0.10397894,-0.2269670
3,0.00173299,0.00100654,0.00147123,0.08082343,0.11429829,0.24383818,-
0.07688582,-0.04244995,0.07352603,0.00331309,0.00442711-
-0.01113397,0.00331337,0.00742900-
-0.00940146,0.07886221-
-0.04244995-
-
0.10690043,0.10397915,-0.01185639,-0.01271862,0.02558442,0.00742897,0.00856605,-0.01
329531,0.04666288,0.11185561,0.07352603,0.10397915,-0.22696703,-0.0017330-
,-0.00100656,0.00147128,0.00940143,0.01329533,-0.01981367,-0.080823

09,-0.11429853,0.24383818,-0.31701629,0.00000327,0.00000023,-0.0086023

0,-0.00042865,0.,-0.00860290,0.00021449,0.00037137,-0.00860286,0.00021

449,-0.00037136,0.34282435,0.00000327,-0.04686849,0.,0.03213837,0.0026

0791,-0.00000002,-0.01606900,-0.00080262,0.00196941,-0.01606904,-0.000

80262,-0.00196939,-0.00000360,0.04586582,0.00000023,0.,-0.04686849,0.0

0000002,0.,-0.00193985,-0.02783250,0.00196974,0.00147128,0.02783250,-0

.00196974,0.00147123,-0.00000025,0.,0.04586582||0.,0.,0.,-0.00005060,0

.00014313,0.,-0.00005061,-0.00007157,-0.00012396,-0.00005061,-0.000071

57,0.00012396,0.00015181,0.,0.||||@

... FOR AFTERWARDS A MAN FINDS PLEASURE IN HIS PAINS,

WHEN HE HAS SUFFERED LONG AND WANDERED FAR.

-- HOMER

Job cpu time: 0 days 0 hours 0 minutes 4.0 seconds.

File lengths (MBytes): RWF= 5 Int= 0 D2E= 0 Chk= 1 Scr= 1

Normal termination of Gaussian 09 at Sun Nov 29 23:31:58 2020.

SUBJECT INDEX

www.ingramcontent.com/pod-product-compliance
Lightning Source LLC
Chambersburg PA
CBHW080019240326
41598CB00075B/425